The Oxbow

Science

S0-BKF-899

THE RED BOOK

THE EXTINCTION CRISIS FACE TO FACE

GENERAL DIRECTION
PATRICIO ROBLES GIL

COORDINATION
RAMÓN PÉREZ GIL

EDITORIAL DIRECTION
ANTONIO BOLÍVAR

TEXTS
AMIE BRÄUTIGAM AND MARTIN D. JENKINS

SPECIAL CONTRIBUTIONS
GEORGE B. RABB, GERARDO CEBALLOS, PAUL R. EHRLICH,
ARTHUR E. BOGAN, AND HOLLY T. DUBLIN

CEMEX

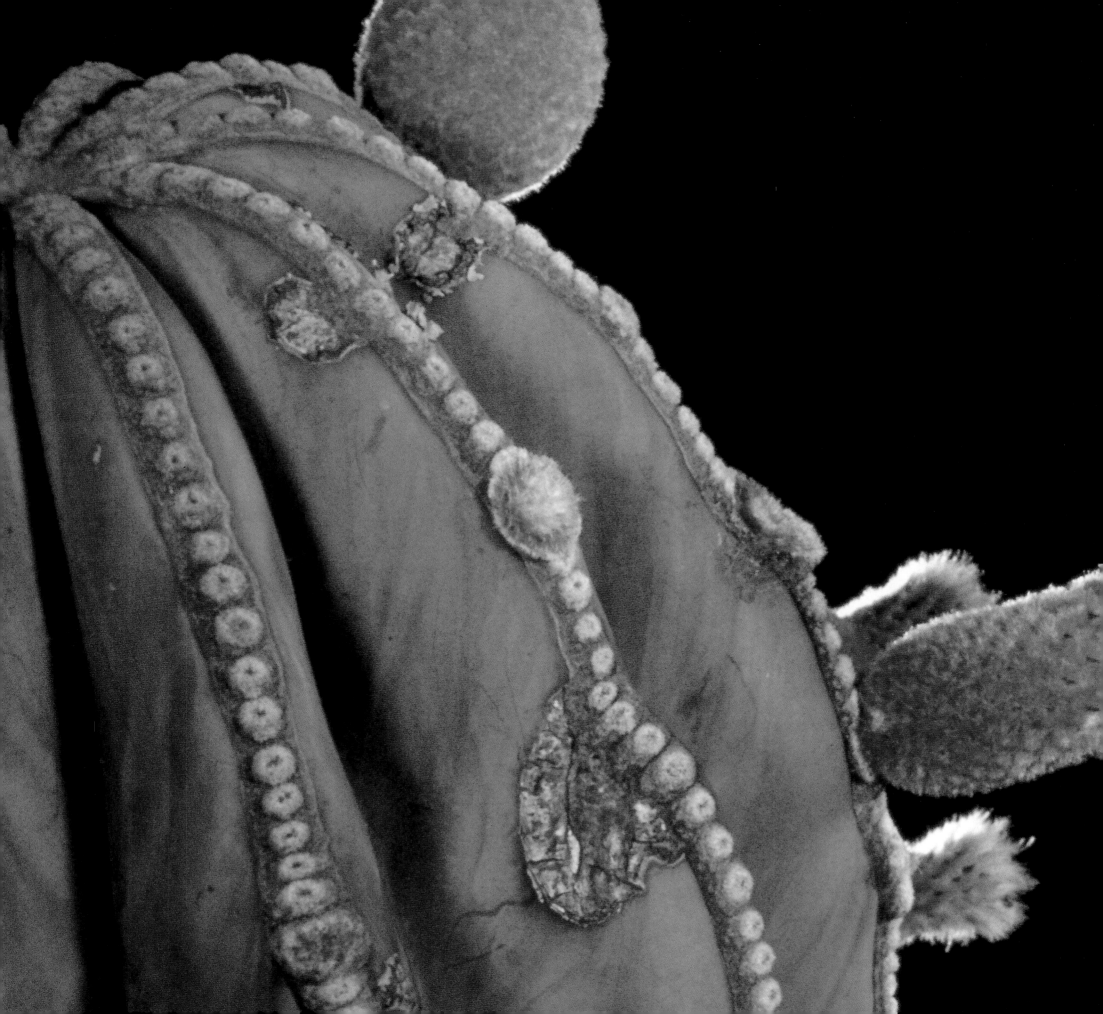

crisis. A turning point in time involving a serious affair having momentous consequences and a great likelihood of leading to an undesirable conclusion.

extinguish. To cause certain beings or entities, such as a life-form, which are disappearing gradually, to cease to exist altogether.

Life on Earth never fails to amaze us: birds that sing with 400 voices; orchids that conceal the scent of countless fragrances; mammals that submerge in the ocean to 500 meters deep; time-honored trees born 40 or 50 centuries ago which are still alive; thousands of plant species that have provided sustenance for civilizations and have also offered a possible solution for cancer... and could they do the same for AIDS?

What is even more surprising is that all the species described by science may not even comprise a tenth of all those inhabiting our planet, which are just waiting to be discovered.

This marvelous universe and all the elements supporting it constitute biodiversity, and there is not a single dictionary that can encompass the full meaning of this term: amazement, beauty, well-being, celebration, leisure, hope, happiness, movement, patrimony, peace, permanence, future, wealth, health, security...

In the next thirty years, our planet will lose one fifth of the species living on it today. And just one species is responsible for this extinction. In the face of this biodiversity crisis, every man and every woman should choose between two paths: the arduous, lengthy process of change towards a sustainable society, or to continue hastily towards the degradation of our natural resources, in which case it is impossible to predict the magnitude of the social, economic, and political problems we shall have to confront.

The challenge is overwhelming, we should act decisively and also immediately. For that reason, with the intention of promoting and strengthening alliances and actions, CEMEX presents *The Red Book. The Extinction Crisis Face-to-Face.*

Thus, we join in the work done by one of the most prestigious international organizations, the World Conservation Union and its Species Survival Commission, which for five decades has promoted the conservation of nature and particularly of biological diversity as an essential element for the advancement of humankind.

The Species Survival Commission is comprised of a group of more than 7 000 renowned scientists, government officials, and conservationists from 188 countries who, on a voluntary basis and for many years, have spoken loud and clear in favor of species conservation.

The information that has been amassed and disseminated is unparalleled; suffice it to mention the impressive number of publications, outstanding among which is the series of books and red lists issued since 1966 documenting the alarming increase in threatened species. These endeavors have been useful in providing advice to governments, international conventions, and conservationist groups throughout the world in order to promote, design, and implement laws to protect species. In the words of Russell Mittermeier, president of Conservation International and chairman of the Primate Specialist Group of the Species Survival Commission: "This process has had enormous catalytic impact, and must be rated as one of the most significant achievements in the history of biodiversity conservation."

Through our publishing efforts spanning nearly a decade, at CEMEX we have been committed to the task of disseminating the world's biological wealth, underscoring the major strategies for conserving it. On this occasion, by issuing *The Red Book. The Extinction Crisis Face-to-Face*, we are striving to be the bearers of that discourse and to invite the international community, governments, conservation organizations, research centers, and society as a whole to join together and confront face-to-face this extinction crisis affecting animal and plant species, assigning it one of the top priorities guaranteeing the long-term permanence of this planet's biodiversity.

Extinction is forever. We have no doubt as to what this means.

CEMEX

The natural and cultural assets of the countries of the world are their most important treasure. It is essential to learn how to manage that wealth in order to fight poverty.

A little over 50 years ago, the founders of the International Union for Conservation of Nature —the IUCN, now known as the World Conservation Union, although it has retained its original acronym— evidenced signs of a most praiseworthy vision by establishing a specialized and unique organization comprised of sovereign states, government agencies, international organizations, and nongovernmental organizations. It has six technical commissions constituting the world's most important network in the field of conservation. These commissions are made up of more than 9 000 professionals and experts. And that vision has become a reality: today, the IUCN participates in countless major conservation initiatives at the local, national, and international level; it unites the efforts made by sectors, interest groups, and nations in the face of the necessity of shaping participatory schemes for resolving issues which threaten the future of all humankind.

The productive sector has joined forces in the challenge of confronting environmental problems, increasingly convinced of the importance of linking economic development with environmental stability. One of the leaders of this sector and of this new way of approaching development is CEMEX, a company which, with foresight and determination, has taken upon itself the task of fostering knowledge on the riches of the natural world and on the need to conserve them.

The books published by CEMEX can hardly go unnoticed. Besides revitalizing our capacity for admiration and amazement, these works have also conveyed a sense of urgency in the face of environmental deterioration and have served as catalysts for a wide range of actions geared to solving crucial problems.

Therefore, it is a source of great satisfaction for me to have the opportunity to express my appreciation and recognition to CEMEX for its farsightedness and leadership.

In this alliance with CEMEX, the IUCN seeks a new opportunity to invite others to join in this great world crusade supporting our planet's survival and the stability of the conditions making life on Earth possible. With this step, we launch what we hope will be a most fruitful and long-term collaborative effort.

The chapters of this book were prepared by members of our Species Survival Commission, one of the most important components of the IUCN and, without a doubt, one of its strongest pillars. Through this endeavor the broad network of volunteers, veritable apostles of conservation who devote time, energy, resources and, in fact, their lives to other living beings, confirm their vocation to serve and their professionalism. One of the most notable achievements of this altruistic group of scientists and experts is the frequent production of "red lists," which are a tool this work evokes and to which it pays tribute.

As a whole, this book is just like a snapshot, a means for contemplating, face-to-face, the drama and complexity of the loss of species with which we share this planet. It allows us to halt before our very eyes, a dynamic process that is mind-boggling. This process has so many facets that fully meeting the challenge it poses seems to be next to impossible. However, it is fair to say that the authors of this book have succeeded in a truly difficult task: that of transmitting this intricate process, accurately termed the "species extinction crisis," in several brief texts written in plain and simple language. At the same time, they have traced rays of hope, incentives for stepping up our efforts so as to reverse current trends and —it is our most fervent wish— on the short term, to demonstrate, as a species, that we deserve to bear our scientific name: *Homo sapiens*.

YOLANDA KAKABADSE
President, The World Conservation Union

The evolving efforts of the IUCN Species Survival Commission (SSC) to identify and track the world's threatened species through the IUCN Red List have played an important role in focusing attention on the sixth extinction wave that we are now experiencing. The Red List has also been a major driver of countless conservation initiatives to save severely endangered species around the world. These advances have included many conservation successes, examples of which are presented in the book we now introduce: *The Red Book. The Extinction Crisis Face-to-Face*. Yet, as SSC member Gerald Kuchling points out, for all the effort —and the successes—, many of these species are still Critically Endangered and require constant, intensive management to maintain them even at a precarious level. The thousands of specialists, scientists, and caring individuals who are part of the SSC can be proud of the increasing awareness and conservation effort that Red Lists have generated. But we are all most sensitive to the fact that much more action is needed.

Throughout the five decades of its history, the Species Survival Commission has focused a great deal of its activity on documenting and raising awareness as to the alarming trend of increasing numbers of species at risk of extinction. The *IUCN Red List of Threatened Species* is the result of an evolution that began when Sir Peter Scott, then Chair of the SSC, initiated the production of the Red Data Book series in the early 1960s. This was a major step forward, as, until that time, the SSC (then known as the Survival Service Commission) had recorded information on the status of species on a series of data cards. The move to a Red Data Book series enabled the SSC to begin to draw public attention to the plight of species, the threats they are facing and, most importantly, the actions that are necessary to preserve species —actions that form the core of the SSC vision, a world that values and conserves present levels of biodiversity, to this day.

From its beginnings as the Red Data Book series, the current *IUCN Red List of Threatened Species* and SSC Red List Programme have grown. The IUCN Red Lists are now based on a systematic assessment of species against precise and quantitative Red List Categories and Criteria, adopted in 1994 and revised in 2000. Concurrently, documentation requirements and taxonomic standards have been set in order to bring greater credibility to the listings, and better analyses of the findings. This more objective assessment process will be undertaken on an annual basis by SSC-appointed taxonomic Red List Authorities, volunteers with expertise in specific species or groups of species and their conservation needs. And we look to further growth, fully documenting the species currently listed, as well as expanding the breadth of species that are evaluated. Plans are under way to assess more amphibians, reptiles, freshwater fishes, invertebrates, and marine species, as well as many more groups of plants. In addition, SSC is pleased to be moving towards assessing broader biodiversity trends and assisting with priority-setting at national levels. This has been made possible with the new categories and criteria, while our capacity to do so is increasing with the development of the Species Information Service (SIS), SSC's global database of species-related information.

We are keenly aware that information alone will not slow or halt this extinction crisis; with the Red List as the guide and catalyst, SSC is engaged in a wide range of activities to prevent extinction and conserve biodiversity. Our Specialist Groups, volunteers dedicated to the study and conservation of species and groups of species, have not only applied their intellectual skills to assessing the status of species, but have also developed strategies and recommendations for action, and worked with conservation agencies and others to see them through. We are pleased to present in this volume a few of these many instances when SSC members and Specialist Groups have brought their knowledge and expertise to bear in preventing the disappearance of a species.

The SSC Programme and Specialist Groups work to provide decision-makers with the knowledge and tools to make sound decisions about species and ecosystems and the people who depend on them. SSC Specialist Groups produce action plans —global strategies, compiled through meticulous analysis of available information and broad consultation, that review and present priorities for conserving groups of species. These include relevant information on the status of the species reviewed and, more importantly, concrete recommendations for action. Aside from charting the course for the SSC and the individual Specialist Groups, these action plans are an important tool for a wide range of agencies and institutions involved in species conservation; finally, they serve as a basis for tracking progress or lack of it. Through the Wildlife Trade Programme, SSC provides advice to national conservation agencies regarding the implementation of trade controls and other safeguards to ensure that trade is not detrimental to the survival of species. The Plants Programme brings its focus to an often-overlooked yet vital component of our global ecosystem. The Invasive Species Specialist Group, Reintroduction Specialist Group, Conservation Breeding Specialist Group, and Declining Amphibian Populations Task Force all address specific threats, issues, or conservation tools, and guide actions which apply that knowledge, whereas the Sustainable Use Specialist Group faces the challenge of helping people balance the resource needs of society with the conservation of wildlife and wild lands so that all benefit in the long term. These are a small sample of the range of activities aimed at achieving our SSC vision.

The SSC and its more than 7 000 volunteers are dedicated to halting the extinction crisis, but this small community will not succeed in isolation. The extinction crisis must be taken up by the greater global community —it needs to be apparent to everyone sharing this planet just what we are all at risk of losing. *The Red Book. The Extinction Crisis Face-to-Face* is one step we are taking to reach out beyond conservation agencies and scientific networks, to bring home the point that this is an issue which touches and implicates all of us. This step could not have been taken without the generous support and collaboration of CEMEX and Patricio Robles Gil, along with the staff of Agupación Sierra Madre. In addition, I must thank Ramón Pérez Gil for championing this book within the SSC family and Amie Bräutigam for undertaking the Herculean task of compiling and editing it over a very short period of time. In addition, I extend my sincere thanks to her coauthor Martin D. Jenkins and the other SSC contributors for filling this volume with the breadth, richness, and personal perspective that make the loss of species so compelling an issue. Of course, all of this would also not have been possible without the SSC members, whose devotion to species is awe-inspiring —as will be seen throughout the images and examples presented here.

The powerful images in this book represent only a tiny portion of the 11 000 or so species currently included in the *IUCN Red List of Threatened Species*, not to mention the hundreds of thousands or more that have not been evaluated, less described or even discovered. But the variety of the images —animals and plants and the places they inhabit— and the sheer splendor that they exhibit are moving testimony of how impoverished our world will be if we let this extinction wave run its course.

DAVID BRACKETT
Chairman, Species Survival Commission, IUCN

INTRODUCTION

What sets worlds in motion
is the interplay of differences, their attractions and repulsions.
Life is plurality, death is uniformity.
OCTAVIO PAZ (1914-1998)

The threats to the continuity of life on this planet are manifold and diverse; in fact, many of them go unobserved. We still know very little about global biodiversity —total figures on species are mere approximations—, and even less about the subtleties of the mechanisms leading to their disappearance.

Humankind responds to short-term natural phenomena that cause immediate damage having enormous repercussions. It would seem that then, our basic values are revived: solidarity and the best facets of human nature. In contrast, when faced with a catastrophe of a greater magnitude and significance —such as the gradual deterioration of the conditions making life on Earth possible—, there is no general, concerted response. It is undeniable that extinction is not an ephemeral occurrence (such as an earthquake), but no one should question the seriousness of its consequences: extinction is irreversible.

At present, the distance that has come to separate humans and the natural world is most evident. Apparently, we have forgotten that we are part of nature and we have lost an awareness of the origins of the resources we use —for example, water— or of our relationships of mutual dependence. The chances for recovering that lost awareness point increasingly towards dissemination and the sensitization of society.

The publishing work that CEMEX has been conducting since 1993 is in keeping with a communication strategy seeking to draw attention to our most pressing environmental issues, as well as to the efforts made by agencies and institutions confronting them from different angles.

Through their inclusion in books, exhibits, videos, and pamphlets, strategies such as these have helped envisage, update, broaden, and improve the environmental policies established in different countries throughout the world; they have enabled us to obtain a great deal of funding for our cause; they have opened up the doors of governmental offices and those of private industry for us and, in this way, have become a model with vast catalyzing potential and crucial results in our attempts to conserve biological diversity.

A little over 50 years ago, a single organization became the leader in the mission of documenting, disseminating, and combatting the threats jeopardizing many animal and plant species on this planet. We are referring to the International Union for Conservation of Nature (IUCN), currently known as the World Conservation Union.

For that reason, we proposed to CEMEX to publish, along with the Species Survival Commission (SSC), *The Red Book. The Extinction Crisis Face-to-Face* in order to make known the important accomplishments of this Commission, which has played a prominent role thanks to its vision, its publications, the work done by its specialist groups, and the influence it has had in contributing towards species conservation.

Many are the reasons why we felt it was important to support, through this book, the IUCN's long trajectory and, in particular, the efforts made by the SSC, one of its six volunteer commissions.

This Commission advocates a modern approach to species survival and an innovative conservation ethic and practice which proposes that conservation can cease to be an economic burden and, rather, become an opportunity.

We set various goals for ourselves: the foremost was to illustrate in a clear, interesting manner the magnitude of life on Earth, its importance for humankind, its present critical state, the work done by the SSC, and the way in which that Commission focuses its efforts on solving these pitfalls; i.e., we intended this book to be much more than a lavish sampling of beautiful, fascinating pictures.

In the process, we came across several problems: we found a tremendous void in terms of information and also images of certain groups of plants and invertebrates, and we also discovered that it is impossible to identify a number of species and, therefore, to determine their current conservation status. Nevertheless, it is startling to learn that, despite that void and with certain exceptions, nearly all the photographs we have included in this book depict species evaluated by the network of SSC scientists and which, once their conservation status had been determined, unfortunately had to be placed in a category indicating they were in trouble.

Yet there were other groups for which we were unable to locate images because up to now, professional photographers have not had any commercial incentive to portray them. To resolve that problem, we endeavored to find photographs taken by specialists who, for the most part, are the only ones that have them. However, we couldn't use many because they were images taken in laboratories or with simulated backgrounds and they just were not what we were looking for.

We also made an attempt to include some of the countless species which play such an important part in the miracle of life and which we virtually do not know either through pictures or verbal descriptions, even though these millions of species of invertebrates, fungi, lower plants, and other life-forms are at least as responsible, if not more so, for the continuity of life on this planet as the plants and animals we so like to admire in nature photos.

However, whatever we have achieved in this book falls very short of reflecting the great biological wealth it aims to capture, namely the very inspiration of this volume: the beauty of life.

Yet this book evinces not only the dedicated, impassioned work of many specialists, but also that of other actors who promote knowledge about species and their conservation: nature photographers, i.e., men and women who —just like scientists— wage their own personal battles and thanks to whom, we are able to gain familiarity with thousands of these life-forms that lack a voice, and yet acquire one through photography and are then able to speak to us face-to-face.

Extinction is an issue that should be as great a concern to society as unemployment or inflation, not only because this topic undoubtedly deserves much more attention but because, as we are well aware, these topics are interrelated. All economies are sustained and are in a position to meet demands and needs when they have access to a steady, abundant supply of raw materials which they convert into goods and services. But let us not forget that nature is the only "manufacturer" of raw materials and that there is no stock comparable, in terms of abundance and diversity, with the millions of plants, animals, and other life-forms living with us here on this planet.

A brief rundown of the elements we need for our daily life highlights this fact: the food we eat, the clothes we wear, the kind of shelter we take, and our very livelihood are all ultimately derived from nature.

During the last few decades, many species have disappeared and yet others have managed to survive thanks to the efforts and commitment made by humans. For example, some twenty years ago, the last remaining individuals of the California condor were captured for the purpose of conducting a captive-breeding program, a project that was totally unprecedented in the history of conservation. More than 25 million dollars have been invested in this program. And although there is still no guarantee that this species will survive, 50 of these majestic birds have now been released into the wild.

When attempting to conserve the diversity of life on Earth, we should not skimp on either efforts or resources. Although we recognize a certain use, exchange, optional, or intrinsic value for each species, we are faced with the distressing necessity of setting priorities for their conservation. But we must be practical. The logic here is quite simple: it is easier to put out a match than a bonfire, and to extinguish a bonfire than a forest fire. It is far more economical to take effective, immediate action towards preventing extinction and thus maintaining the greatest possible species diversity; it is wiser than to wait until these species have virtually disappeared from the face of the Earth and then to invest exorbitant sums of money to try to keep them from going extinct.

Life is a force that sets the world in motion and holds it there. We now know that the continuity of that movement is at risk. Species extinction, like the tip of an iceberg, attests to that fact: it is a critical and alarming global problem. Species of plants, animals, and other life-forms are being lost even before science has had an opportunity to discover them, much less describe them; these species, which have great potential for improving health and nutrition all over the world, are vanishing forever. With their disappearance, we relinquish links with our past and deprive ourselves of the possibility of a diverse, fruitful life in the present, as well as alternatives for the future.

PATRICIO ROBLES GIL
President, Agrupación Sierra Madre

RAMÓN PÉREZ GIL
*Executive Committee Member,
Species Survival Commission/IUCN*

The following text offers an awesome argument for admirable greed. Simply put, admirable greed to continue to cherish all of the remarkable and fascinating diversity of life that shares the Earth with us. The sheer wonder of this diversity is celebrated in the astonishing images in this book and in two previous, outstanding books on human and biological diversity and on the concentrations of diversity in certain places on the planet, so-called hotspots. The focus in this book is on the progressive loss of the natural wealth of the living world and the need to act to save as much as we can.

Why should we so act? In my view, we should act to manifest all our own innate characteristics as a species and to fully realize the ideals that many people embrace as citizens of the world. People generally wish to be complete persons, and exercising admirable greed for the diversity of life will help us individually reach this goal. I know that this may seem far-fetched reasoning, but consider some of the distinctive behavioral attributes of our species.

For one, appreciation of beauty is a feature of human behavior that we long to fulfill. Beauty is pleasing sensation or perception in the eye and other senses of the beholder. What does the beholder see that evokes pleasurable feeling? Various features of individual animals, plants, and people seem to qualify —symmetry, col-

oration, structural intricacy. It happens that, among the members of any particular species, the various expressions of such characteristics may also indicate differences in likelihood of survival. Thus, in a way, maintaining beauty equates with maintaining life-styles.

While other animals may also appreciate symmetry, it seems that people alone extend the feeling for beauty to the integrity and completeness of others' bodies and of assemblies of other creatures. We seem to be internally pleasurably aware of the completeness of a community of living things. Some people also take delight in the sublimity of appreciation of the very diversity of groups of related organisms.

The diversity, the variety of living things allows us to exercise another universal feature of our behavior —our curiosity. We seem compelled to discover novelty in new things, relationships, and phenomena, whether in the physical realm about us or in the living world, including our own bodies. Again, the bounty of nature's diversity assures us that there is no need to deprive ourselves of a boundless source for satisfaction of our curiosity.

An extension of our curiosity is our fascination with how things work. Problem-solving tickles us so much that we invent puzzles of all kinds and there is a whole abstract field of puzzles that we call math-

ematics. Again, look at what nature offers us in the thousands of ways in which different creatures make a living, and the multitude of variations on the themes of different life-styles. We have only begun to explore this diversity. The puzzles include coping for the basic essentials needed for existence and propagation of one's kind, and how to exist in diverse, challenging and changing environments, as well as in viable relationships with other inhabitants of the same environments.

Among other behaviors, we collect things and we name things. Although we do not fully understand the basic pleasures that come to us from these activities, they are part of our makeup. In the diversity of life, nature has given us a wonderful gift that can endlessly meet our urge to collect and name. While we have named more than a million kinds of living things, it is conservatively estimated that there are at least ten times more. And, of course, each of these creatures has numerous attributes that we can determine and categorize to our heart's content.

The foregoing may appear to be a selfish and foolish argument for ensuring the maintenance of the diversity of life. Many would rather have fellow people act on moral or rational grounds to ensure that the panoply of life continues to exist in a natural state alongside us. A combinatorial argument is that it is unethical for us to deny the pleasures of the living world to fellow people or future generations.

Unfortunately, conservation advocacy because of ethical transgression or of lost ecological and economic values has not diminished the accelerated, unnatural rate of loss of species in recent times around the globe. The impact that our species is making on environments and their living constituents has been compared to the catastrophic episodes of extinction of life in the Earth's past. We humans are the equivalent of the external agents of large asteroid impacts on Earth that appear to have fatally changed the environments for the dinosaurs and other forms of life prevalent in the past.

Whether one is motivated by selfishness or altruism to help maintain the diversity of life, there is much that can be done individually and collectively. It is certainly not necessary to know all the details of the biology of a plant or animal species to avoid its possible or probable extinction. What this task really calls for is commitment and determined effort to change the perilous conditions that we have imposed directly or indirectly on species and their environments. We have encouraging examples of the saving of species of animals threatened with extinction —bison, bald eagle, American alligator, Arabian oryx, Asian rhino, and golden lion tamarin, to name a charismatic few.

While unsustainable hunting or harvesting, habitat destruction or degradation, and immediate chemical pollution are direct threats that have been dealt with in saving the species just named, other situations are far from simple. For instance, the global decimations of coral reefs and frog pop-

ulations probably represent combination effects of detrimental chemical compounds. In the case of the frogs, the specific causative agent of sudden population die-offs and extinctions of some wonderful species is an aquatic chytrid fungus. It is likely that this fungus and frogs have coexisted for hundreds of millions of years. What has changed for them? It may be direct effects of man-made chemicals such as the so-called POPs, or synergies or interactions among various chemicals in the atmosphere or on the ground or in the water. Obviously, a puzzle that calls for solution if we care. Fortunately, there are scientists and volunteers who *do* care and are tackling this puzzle. The ultimate solution, of course, may require substantial changes in the conduct of major commercial enterprises and in this realm we will need much more widespread, caring concern.

To conclude, this book makes plain the losses of biological diversity that we are about to cause if we do not employ countering conservation strategies and approaches worldwide. While there have been great initiatives and special projects conducted by governmental agencies and by wonderful volunteers in IUCN's Species Survival Commission and in other organizations, all people need to be concerned and involved. The images and stories in this volume are indeed aimed at encouraging all humans to recognize what we are losing in beauty and environmental integrity if we do not change our individual and societal behaviors. Everyone can help effect positive change in policy and practice at all levels of human society.

From reading and viewing this mar-

velous volume, I hope that other people will come to share the awe that I feel for our diverse companions on planet Earth. And I hope that this awe will, in turn, inspire people to act to help save not only the endangered species of the Red List, but all species everywhere.

GEORGE B. RABB
Director, Brookfield Zoo
and Chicago Zoological Society

MOTHER NATURE'S INGENUITY: THE DIVERSITY OF LIFE ON EARTH

For many of us, the most remarkable feature of life on Earth is its endless, extraordinary, overwhelming variety. The enormous range of living organisms —in form, life history, and adaptation to their environment— takes us from egg-laying mammals to trees that live more than five thousand years. There are cells that thrive in boiling sulphur springs where any other creature would perish in an instant and penguins that incubate their eggs for four months without feeding in temperatures down to –50°C. There are corals that glow in the dark, ants that farm fungi, and fishes that electrocute their prey. Virtually every species on the planet has a physical, biological or ecological peculiarity; marvels of creation, they excite, amuse, inspire, and bewilder us. Making sense of this diversity is one of the most satisfying, and frustrating, undertakings in science. Conserving it —in the face of so many pressures and so many unknowns— is one of the greatest challenges facing humankind today.

Species: The fundamental currency of biodiversity

Species are the most important building blocks of biological diversity and, thus, a major focus of efforts to understand and conserve our natural world. Although the importance of species is universally acknowledged, there is no firm agreement as to how they should be defined, or what constitutes a species in the real world. Uncertainty about actual species and their status, as well as overall species numbers, has major implications for conservation.

In principle, a species consists of a population of organisms which can actually or potentially interbreed (and produce fertile offspring) but can not breed with any other population. Identifying such populations is easier in theory than in practice. It is true that there are many unequivocal species, clearly distinguishable from all other organisms, such as the giant anteater *Myrmecophaga tridactyla*, duck-billed platypus *Ornithorhynchus anatinus* or maidenhair tree *Ginkgo biloba*. Often, however, the distinction between one species and another is much less evident, for a variety of reasons. Most importantly, those who classify species usually do so on the basis of physical characteristics or morphology rather than through a study of living organisms reproducing in their natural environment. While this usually provides a good guide, it can be misleading: morphologically very similar organisms may belong to quite separate species while, especially in plants, individuals of a single species may look very different depending on their growing conditions. Moreover, taxonomists often disagree amongst themselves as to how different two individuals have to be in

to page 38 →

34

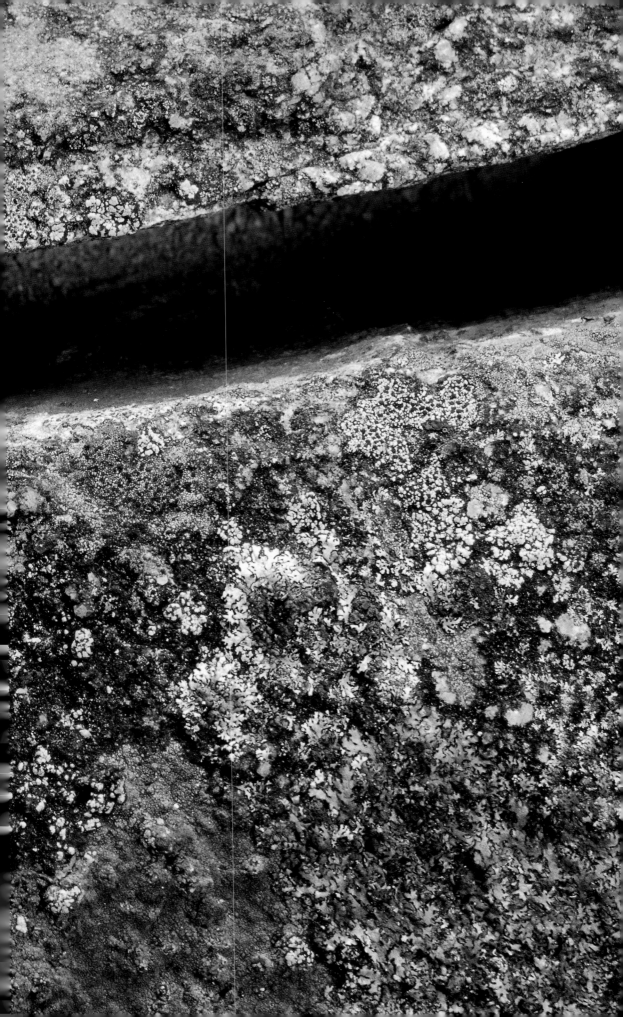

order to be placed in separate species. The use of different classification systems can give rise to a proliferation of different names, or synonyms, for the same species, causing further confusion. These inconsistencies are compounded by the constant revisions that result from new discoveries or reanalysis of existing genetic material. While recent developments in the description of species based on direct analysis of chromosomal material are resolving uncertainties about some species, they are giving rise to questions about others.

This state of affairs —a tentative, ever-changing knowledge base— is one of the everyday handicaps of the conservation endeavor and is unlikely to stabilize in our lifetime.

The diversity of species

The great number and variety of species are organized by scientists in a taxonomic hierarchy. Despite ongoing debates and revisions, there is broad general agreement on many of the major categories comprising this hierarchy. At the very basic level, there is a divide between those organisms that have a cell with a nucleus containing most of the genetic material in the cell, and those that do not. The latter are the ubiquitous, usually microscopic bacteria among whose number were almost certainly the first living organisms on Earth; fossil remains from nearly 4 billion years ago look little different from some existing bacterial forms. Those organisms that have cells with nuclei include all other forms of life, from amoebae to whales. This latter group, known as eucaryotes, is itself divided into four major groups or **Kingdoms**: animals, plants, fungi, and the protoctists or protists, an assemblage of primarily microscopic organisms that includes many forms of algae, including the distinctly nonmicroscopic seaweeds (previously considered plants) and the group formerly known as protozoa, amongst which are many tiny free-living aquatic organisms, as well as a number of important disease-causing parasites such as the malarial agent *Plasmodium*.

Further down the taxonomic hierarchy, **Phyla** represent the major types of body plan found in nature. Most of the organisms with which we are familiar in everyday life fit into relatively few of these: mammals, birds, reptiles, amphibians, and fishes all belong to a single phylum, the Craniata or Vertebrata. Molluscs —snails, slugs, clams, and other seashells, squid, and octopus— comprise another. Earthworms form part of the phylum Annelida, while flowering plants constitute the Anthophyta and conifers, the Coniferophyta. For every phylum containing at least some well-known species, there are sev-

eral obscure ones, including the Sipunculids, comprising 150 species of misshapen marine worms of up to 50 cm long, and the enigmatic Placozoa, which consists of one tiny species discovered on the glass of a seawater aquarium in the nineteenth century and which seems to be the simplest of all living animals. Phyla are generally subdivided into a number of **Classes**, which include the mammals (Mammalia), birds (Aves), reptiles (Reptilia), and amphibians (Amphibia), and four separate classes of fishes. Classes themselves are further broken down into a number of finer divisions, the smallest component of which are species (and subspecies).

While few would suggest that any one species or group of species is fundamentally more important than any other, the degree of distinctiveness, based on evolutionary history, can be an important consideration for the conservation of biodiversity. Examples of species with few, or no, close living relatives abound. The Chinese maidenhair tree *Ginkgo biloba* is the only living member of its phylum, the Ginkgophyta. The two species of tuatara, *Sphenodon guntheri* and *Sphenodon punctatus*, "living fossils" found only on a number of islands in New Zealand, are the sole surviving members of the order Rhynchocephalia, a group of reptiles that was common between 225 and 125 million years ago. These unique organisms —isolated branches on the tree of life— are of such great scientific interest that their loss would be a particularly severe blow.

How many species are there?

The number and distribution of species that exist today are a product of more than 3.5 billion years of evolution involving speciation, migration, extinction and, more recently, human influences. One of the most persistent debates amongst taxonomists, conservation biologists, and other scientists revolves around estimates of the total number of species in existence. These estimates range from 7 to 20 million, but the current best working estimate is considered by many experts to be between 13 and 14 million species, of which only around 1.75 million species have been described. That the great majority of species are still undescribed is inferred largely from the high rate of discovery of new species from tropical rainforests and the seabed of the world's continental slopes.

The completeness of our knowledge of different species groups varies greatly, as does the rate of discovery of new species (see Table 1). By far the best-known major taxonomic groups are mammals and birds, but even with these, revisions of species and species checklists are constantly being made, with new species being named and others being declared invalid. More rarely —but not infrequently—, genuinely new species are being discovered. Descriptions of new mammal species are currently published at the rate of around twenty per year; most are members of the more numerous and least well-studied groups of smaller animals, particularly rodents, insectivores, and bats. However, occasionally something much larger and more surprising appears. Perhaps the most dramatic recent mammalian discovery has been of the Saola or Vu Quang ox *Pseudoryx nghetinhensis*, a distinctive-looking member of the antelope family or Bovidae, which was first recorded by scientists in the 1990s in the Annamese highland regions of Viet Nam and the Lao People's Democratic Republic.

There is greater uncertainty about species comprising the other vertebrate groups. No complete checklist of reptile species exists, for example, although there are lists of all snakes, tortoises, and crocodilians, and new species of reptiles are being discovered on a regular basis. Although a complete global checklist has been compiled for the amphibians, this group, too, remains much less well known than the mammals or birds, and new discoveries are routinely being made. Amongst fishes, our level of ignorance is even greater: a checklist of species is in progress, with the first three volumes published in 1998, but it is widely accepted that a considerable number of species remain to be discovered, particularly in the deep sea and in the great river basins of the tropics, most importantly the Amazon.

Even with this uncertainty, and the steady flow of new discoveries, there is no doubt that the great majority of vertebrate species are known to us. Roughly 52 000 species are currently described (see Table 2) out of a predicted total of at least 55 000, with most of the currently unknown species being fishes.

Amongst invertebrate animals —those in the remaining 36-odd phyla— the situation could not be more different: even the most conservative estimates suggest that no more than one out of every five invertebrate species has been described, and there has been serious speculation that unknown species may outnumber known species by twenty or even one hundred to one.

Our knowledge of plant species lies somewhere between that of vertebrates and invertebrates. A far higher percentage of plant species is believed to have been described than of invertebrates, but it is much more difficult to compile full lists of families, orders or other higher taxonomic groups, and estimates of the total number of described species remain far more approximate. One reason for this is that botanists tend to work on a geographical basis, compiling checklists (or floras, as they are known) of particular countries or regions. Floras from adjacent countries or regions often treat populations of plants that occur in both areas differently, leading to a proliferation of synonyms, so that without detailed study it is unclear just how many plant species there really are. This problem is compounded by the fact that it is genuinely much more difficult to decide what constitutes a species in plants than in animals, largely because of the very varying ways that plants reproduce.

The variability of species

It is axiomatic in biology that no two species are exactly alike. The differences between them extend beyond differences in size and appearance to take in all aspects of their biology and ecology. Of crucial importance in conservation are differences in the rates at which species reproduce. Many microorganisms can go through several generations in the course of a single day, while some large plants and animals may take decades before they are mature enough to reproduce. The number of offspring an individual produces at once is also hugely variable: the females of a number of large mammals, such as elephants, rhinos, and sea cows, produce just one young every three to five years, while some fishes, such as sturgeons and tunas, may lay literally millions of eggs at once. Evidently, reproductive strategies greatly influence the resilience of populations in the face of threats. Species that reach reproductive age quickly, or produce large numbers of offspring, or both, can clearly recover much more rapidly from population setbacks than late-maturing species with low fertility. It is scarcely surprising that a high proportion of the latter feature in threatened species lists.

Similarly, the capacity of species to adapt to different environments is enormously variable. At one extreme are some parasites that can use only one other species as their host: the pygmy-hog sucking louse *Haematopinus oliveri*, for example, has only ever been recorded on the Critically Endangered pygmy hog *Sus salvanius* of northeast India, and is therefore considered as threatened as its host. In contrast are humans themselves, and those wild species that accompany them wherever they go in the world, such as rats, mice, and cockroaches. The adaptability of these species seems to show few bounds.

Just as species vary markedly in morphology, biology, and ecology, so do the ranges over which they occur. Some species are found over enormous areas: the barn owl *Tyto alba* breeds in no fewer than

to page *50* →

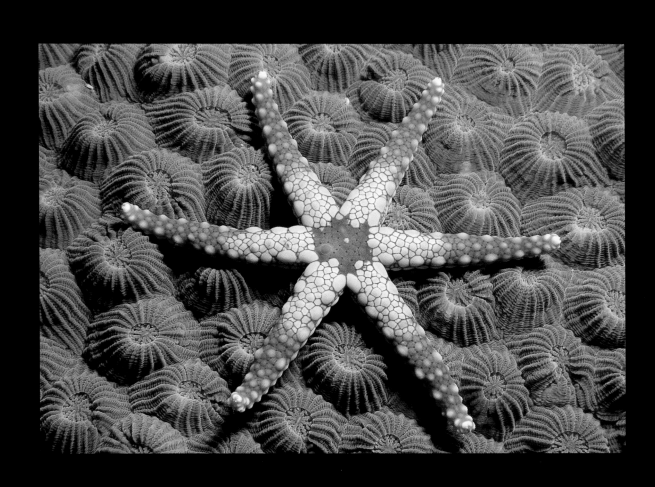

47

150 countries across the globe, while the bracken fern *Pteridium aquilinum* is a ubiquitous feature of upland areas on five continents. The blue whale *Ballaenoptera musculus* and leatherback turtle *Dermochelys coriacea* are just two of the many marine species that travel across huge expanses of the world's oceans. In contrast, other species have minute known ranges, encompassing perhaps a single desert spring, as in the case of the desert pupfish *Cyprinodon diabolis*, whose entire world population lives in an area of some 20 square meters in one limestone pool in Nevada, U.S.A., the smallest known range of any living vertebrate.

Species with ranges that are naturally restricted to a specific geographical unit such as an island, lake or mountaintop, or a geopolitical unit such as a country, are referred to as "endemic" to that area. Endemic species typically have small ranges and form a high proportion of globally threatened species. Increasingly, species' natural ranges are being greatly affected by human activities; a very large number of species have experienced serious reductions in range, while a smaller number have expanded their range. In addition, many species may occur over wide areas but are dependent on highly-specific habitats such as a particular vegetation type (for example, bamboo in the case of the giant panda *Ailuropoda melanoleuca*), a specific altitude or, in the case of aquatic habitats, a specific type of substrate or degree of water flow. Thus, the actual area that a species occupies may be much smaller than at first appears. This type of endemism —in the form of habitat specificity— also has important implications for extinction risk assessment: species with a high degree of habitat specificity are more at risk than those that live in a range of habitat types.

The global distribution of species

The riches of the living world —in terms of numbers of species— are very unevenly distributed over the planet. Most fundamentally, there are profound differences in diversity between the three major biomes that constitute the biosphere: the seas, fresh waters, and land. One major pattern, however, applies to all three: the extraordinary diversity of the tropics compared with the cooler parts of the Earth. One square kilometer of Malaysian rainforest might contain over 1 000 tree species, more than the whole of temperate North America. Over three times as many bird species —just under 1 400— breed in Ecuador than do in Canada, a country over twenty times the size. With a small number of exceptions —which include seals, seabirds,

and some groups of lichens and liverworts—, this overall trend appears virtually universal. The reasons for this have been hotly debated by ecologists. Whatever the cause, it is clearly one of the most fundamental patterns in nature.

Diversity in the seas

Marine ecosystems form by far the largest part of the biosphere. The seas cover over seventy percent of the surface of the planet, have an average depth of nearly 4 kilometers, and hold an estimated 1.37 billion cubic kilometers of water. Virtually the whole of this vast realm supports life, although very large parts of it appear to be almost barren.

All evidence to date indicates that life evolved in the sea and appears to have been confined there for at least the first three quarters of its existence on Earth. As a consequence, considerably more phyla are present in the seas than on land or in fresh waters —amongst eucaryotes, approximately 60 are found there, compared with 40 or so each on land and in fresh water. With animals, this discrepancy is even more marked: 36 of 38 phyla have marine representatives, and of these, approximately 23 are believed to be confined to marine habitats.

In terms of known numbers of species, however, the situation is reversed. There are believed to be only around a quarter of a million described marine species (see Table 3), compared with an estimated 1.5 million terrestrial ones. This discrepancy holds even amongst the best-known groups. Thus, the total known number of marine vertebrates is estimated at *ca.* 15 000 (virtually all fishes), compared with *ca.* 22 000 terrestrial vertebrate species, *ca.* 10 000 freshwater forms, and some 5 000 amphibians, which occur by and large both in fresh water and on land. It has been argued that the relative poverty of the seas in terms of species is merely a reflection of our greater ignorance of marine ecosystems: although this situation is changing rapidly with advances in technology, great expanses of sea remain virtually unexplored. It is true that exploration of the deep ocean and the ocean floor is uncovering many new species, particularly of tiny invertebrate animals and protoctists of various forms. The rate of discovery has led some marine biologists to speculate that diversity in some areas might rival that found in the most diverse terrestrial ecosystems. Most experts, however, believe that this is unlikely to be the case and that, although there may conceivably be as many as half a million undescribed species in these environments, there are still too few to rival terrestrial diversity.

As might be expected, not only are overall totals of species diversity very different in the sea from on land, but the same is true of the relative abundance of the different major groups. While on land one phylum —the so-called Mandibulata (the insects and their relatives)— vastly outnumbers all others, in the sea there does not appear to be any single group that dominates to this degree. The most diverse known groups are the molluscs (*ca.* 75 000 described species) and the crustaceans —the crabs, lobsters, shrimps, prawns, and their relatives— with around 35 000 known species. Other notably diverse groups include the echinoderms (the starfishes and sea urchins), the sponges or Porifera, and the corals, sea anemones, and jellyfishes, which together comprise the phylum Cnidaria. As on land, a significant but poorly-known component of diversity is made up of worm-like creatures of various sorts: flatworms or platyhelminths, roundworms or nematodes, and segmented worms in the phylum Annelida.

One very striking difference between the sea and the land is the very low diversity of plants in the sea. Only 60 or so species of plant, the sea grasses, can be considered truly marine. Sea grasses occur over extensive areas in coastal zones around the world to depths of *ca.* 60 m; although not diverse, they constitute an important habitat for many marine species, including sirenians (dugongs and manatees) and many fishes, which are dependent on them for grazing and as nursery areas for their young. The loss of seagrass beds to development, pollution, and other impacts attributable to humans has had serious negative consequences for these species.

On land, green plants are largely responsible for the vital role of photosynthesis: the trapping of the sun's energy to transform carbon dioxide and water into both the substance of living things and the energy supply that nonphotosynthesizing organisms (including all animals) can use. In the sea, this function is carried out primarily by various forms of bacteria and a range of protoctists, including algae and diatoms. Many of these marine photosynthesizers are free-floating (or planktonic), microscopic forms. A number, however, may form relatively large or sometimes very large organisms that can be regarded as the marine equivalent of plants. These are the seaweeds, familiar from coastlines the world over. Some 6 000 species of seaweed are known, from three protoctist phyla: the Phaeophyta or brown algae, the Rhodophyta or red algae, and the Chlorophyta or green algae.

Of all marine habitats, the richest in species are undoubtedly the coral reefs that form a band around the Earth's tropics. These shallow, warmwater communities support an amazing array of species of algae, invertebrates, and fishes: one medium-sized reef in the Indo-Pacific could easily be home to 500 species of fishes alone, ranging in size from tiny, centimeter-long gobies to five-meter sharks. Sometimes referred to as the rainforests of the sea, the world's coral reefs, like their terrestrial counterparts, are under severe stress from a variety of factors and, therefore, are considered threatened marine habitats.

Diversity on land

On land, almost as striking as the relationship of diversity to temperature is the correlation with moisture. All other things being equal, the damper an area is the more species are likely to occur there. Thus, unsurprisingly, extreme desert regions, whether hot, such as the Sahara or Skeleton Coast of Namibia, or cold, such as much of the high Tibetan Plateau, generally support few species. This paucity of species is also a feature of polar regions and the world's highest mountains which, although they have ice in abundance, have little or no water in liquid form available, and therefore have the climatic characteristics of deserts. While such areas may have low overall diversity compared with many other regions, the species that *do* occur there often show remarkable adaptations to these extreme environments.

Away from deserts, broad patterns of diversity differ somewhat between plants and animals. Animal diversity increases steadily from polar regions to the tropics, reaching its highest manifestation in the rainforests that straddle the equator around the world. Plants, in contrast, show unexpectedly high levels of diversity at more temperate latitudes, specifically in areas with a Mediterranean climate characterized by cool, wet winters and warm, dry summers. Such areas are found in five separate parts of the world: the Mediterranean Basin itself, coastal California, central Chile, the Cape Province of South Africa, and southwest and south Australia. The largest one —the Mediterranean Basin itself— is home to an estimated 25 000 species of plants, approximately one tenth of the global total. In terms of numbers of species per unit area, the richest one is the Cape Province of South Africa, where 8 500 plant species have been recorded in an area of less than 100 000 km^2; *ca.* 70% of these are endemic, or unique, to the region.

Despite the high levels of plant diversity in Mediterranean-type ecosystems, the most diverse ecosystems on Earth are incontestably

to page *62* →

tropical rainforests. It is impossible to estimate the total number of species occurring in these ecosystems —hugely divergent estimates exist of the number of undescribed insects and other species to be found there— and, thus, the proportion of the world's species that they harbor. However, many experts believe that they probably hold between 60% and 80% of the world's species in an area covering a mere 7% of its land surface.

While the great continental rainforest regions of Africa, Asia, and Latin America are the repository of most of the world's species, other regions are also extremely important in terms of their contribution to global diversity. These are areas that are particularly rich in endemic species. Noteworthy amongst them are tropical and subtropical islands, particularly the larger ones such as Madagascar, the Philippines, New Caledonia, and New Guinea. These regions tend to have somewhat fewer species overall than equivalent continental areas. However, a high or very high proportion of these are endemic and often have very restricted ranges. In addition to being found nowhere else, many of them are remarkable in their evolutionary history and thus of great importance to global biodiversity: the lemurs of Madagascar, New Caledonia's extraordinary kagu *Rhynochetos jubatus*, and New Guinea's birds of paradise are amongst the world's most distinct living organisms.

Diversity in fresh waters

Inland water ecosystems account for less than one hundredth of one percent of the world's supply of water, the remainder being found in the seas, as polar ice, or in groundwater. They also cover less than 1% of the surface of the globe. However, this relatively tiny portion of the biosphere features a huge range of habitats and ecosystems, from ephemeral pools scarcely larger than puddles to vast rivers, lakes, and marshes that may extend over thousands of square kilometers, each supporting its own distinctive suite of species.

Much of freshwater biodiversity remains poorly known, and it is difficult even to collate the number of described species that occur in fresh waters, let alone assess how many unknown species might occur there (see Table 4). It is evident, however, that as in marine environments, the vast majority of freshwater species are invertebrate animals; vertebrates and plants represent a relatively small minority.

Given their small area, fresh waters are disproportionately rich in species. Around one quarter of all known mollusc species and 40%

of all known fish species occur in freshwater habitats, suggesting that there is one fish species for every 15 km^3 of fresh water compared with one fish per 100 000 km^3 of seawater. A similar imbalance applies within freshwater ecosystems with rivers, which contain around one hundredth of the volume of fresh water found in lakes, harboring many more species per unit volume of water. While rivers contain much of the world's freshwater fish diversity, high numbers of endemic fishes and other organisms occur in ancient lakes such as Lake Victoria and the other Great Rift lakes of Africa, and Lake Baikal in Russia, as well as springs in arid lands and caves.

As on land and in the seas, there is a marked latitudinal gradient in diversity, with tropical fresh waters harboring far more species than equivalent temperate or polar areas. Hence, the most diverse freshwater ecosystems on Earth are large tropical rivers, notably the Amazon in South America, Zaire/Congo in Africa, and Mekong in Southeast Asia, and lakes such as Lake Malawi, Lake Tanganyika, and Lake Victoria. All these areas remain incompletely explored biologically, with parts of the Amazon —by far the largest river system on Earth— in particular yielding new species of fish and other animals on a regular basis.

The importance of species

It is not simply as objects of study or curiosity that species contribute to our lives. Species have enormous aesthetic and spiritual value in cultures all across the world, as well as being of symbolic and ritualistic value, and of great economic importance. Individually at times but more often in concert with others, they serve vital roles in the ecosystems that we inhabit or on which our life and livelihoods depend. Hence, the impoverishment of biodiversity through accelerating species loss has enormous consequences for our physical and psychological well-being.

At the most fundamental level, it is living organisms that keep the planet habitable. Without plants, algae, and blue-green bacteria performing the daily miracle of photosynthesis there would be no oxygen in the atmosphere to breathe and nothing for the great majority of species to feed on. Other specialized organisms fix nitrogen —another element fundamental to life— and yet others decompose waste products and dead bodies, keeping vital nutrients recycling through the biosphere.

At a higher level, many plants depend on a range of animals to complete their life cycles. Thousands of species are pollinated by flying insects or birds, some by just a single species, or rely on fruit-eating animals to disperse their seeds; without them, populations may fail to regenerate, shrinking in range and becoming more vulnerable to extinction. Within the animal world, predators and parasites may help maintain stability in ecosystems by ensuring that no one species of herbivore becomes completely dominant.

Although we have discovered much about the roles of primary producers, decomposers, and other kinds of organisms, in reality we still do not fully understand the functioning of even the simplest of natural systems, let alone the bewilderingly complex tropical forests and coral reefs in which so many of the world's species reside. What is becoming clear, however, is that these systems are all held together by intricate, interconnecting webs in which diversity at all levels, from the simplest photosynthesizing organisms to the topmost predators, plays a part. Just how much diversity is needed remains an open question. The mere fact that species can and *do* become extinct —as is discussed in detail in the following chapter— indicates that not all species are always absolutely necessary for ecosystems to continue functioning. Yet it is also clear that ecosystems can not sustain the loss of species indefinitely. Sooner or later, some kind of collapse is inevitable.

The value of species to humans

It is self-evident that no single species on the planet can possibly survive alone. That, of course, applies to humans as much as to any other organism, although it is a fact that we often appear to forget. Our survival depends absolutely on the continuing functioning of the biosphere, which itself is dependent on the activities of a host of other organisms. From this perspective, it seems almost superfluous to try to place a value on ecosystems and species —they are literally beyond price. However, an innovative scientific paper published in 1997 *did* try to do just this. A team of experts estimated that the monetary value of the goods and services provided by natural ecosystems amounted to around 33 trillion dollars (US$ 33×10^{12}) per year, this being nearly twice the global economy resulting from human activities. Most of this was in the form of services currently outside the market system, such as gas regulation, waste treatment, and nutrient recycling, with around two thirds contributed by marine —primarily coastal— systems, and the remainder from terrestrial and freshwater systems, mainly from forests (US$4.7 trillion) and wetlands (US$4.9 trillion).

In addition to these indirect uses —generally grouped as "ecosystem services"—, very many species are also of immense actual or potential direct value to humans, primarily as sources of food, medicines, fuel, and building materials.

Perhaps the most direct and fundamental use that humans put other species to is as food. All our domestic livestock and cultivated crops are ultimately derived from wild species. For some, the process of domestication began over ten thousand years ago, when agriculture developed at the end of the last Ice Age, while others are much more recent additions to our stock. Although there are literally tens of thousands of edible plant species, surprisingly few —around 200— have been domesticated for food production and of these, only a handful are of major economic importance globally, with just a dozen crops providing three quarters of the calorie intake of the world's human population. Wild relatives of these and other crops are potentially invaluable in breeding programs, offering the chance of introducing important traits such as disease-resistance without recourse to complex and sometimes controversial new technologies.

Aside from the potential genetic value of wild relatives of domestic crops and livestock, a great deal of food is still harvested directly from the wild. Globally, by far the most important source of wild foods comes from the world's fisheries, both marine and freshwater. Currently, around 100 million tons of aquatic organisms are harvested from the wild every year, nine tenths of this from the seas, the remainder from inland waters. This harvest, which has grown fivefold in the past half-century, represents a vital contribution to world food security. It also involves a huge range of organisms —well over a thousand species of fishes, molluscs, and crustaceans, and also more obscure forms such as sea cucumbers and sea urchins— that are regularly taken. Unfortunately, the majority of the world's accessible fish stocks are now being exploited at levels beyond anything that they can sustain. As a result, many once bountiful stocks have collapsed in recent years, with disastrous local consequences for aquatic ecosystems and the humans that depend on them.

Although globally less important than fisheries, other forms of wild food provide an indispensable resource to people in many parts of the world, particularly the rural poor. In parts of Africa and Asia, bushmeat provides the majority of protein in many rural households, while in dry regions, wild tubers and roots can literally be lifesavers in times of drought or famine.

to page *74* →

Wild plants, animals, and other organisms also still play a pivotal role in meeting the health care needs of humans. In the 1980s, the World Health Organization estimated that 80% of the population in developing countries relied on traditional medicines on a day-to-day basis. Even in the developed world, where most conventional drugs are now synthesized in laboratories, the majority of these contain compounds either extracted from a range of wild organisms or copied from them. There is also a growing interest in developed countries in traditional alternatives to modern medicine; the vast majority of botanical material used for this still comes from wild sources. The range of species used in medicines worldwide is remarkable —perhaps somewhere between 10 000 and 20 000 plants and scores of animals. Unfortunately, where not adequately controlled, collection for medicines can pose a serious threat to the organisms involved. In East and Southeast Asia, the burgeoning trade in animals such as pangolins and freshwater turtles used in traditional Chinese medicine has reduced populations of several species to critical levels and threatens to do the same to many more.

As well as providing these fundamental necessities, the diversity of nature also fulfills a less tangible, but perhaps ultimately equally important role, in meeting our emotional, spiritual, and aesthetic needs. All around the world, people cultivate a vast range of ornamental plants and keep pets or companion animals for no discernible practical purpose. They do it simply because they derive enormous pleasure from having these manifestations of the natural world in close proximity. Across the globe, a growing number of people with leisure time available to them use that time to bring themselves closer to nature through hiking, bird-watching, fishing, and a host of other activities. They are manifesting something that perhaps we are all inwardly aware of: that we cannot, ultimately, separate ourselves from the natural world of which we are an intimate part, and that if we continue to pretend that we can, the consequences for us, as well as for the other species with which we share the planet, are likely to be disastrous.

The IUCN Red List of Threatened Species:
Towards a global system for monitoring trends in biodiversity

While there is growing awareness of the importance of biodiversity for our continued well-being, and mounting concern about the current loss of biodiversity, the world still lacks a functioning system for monitoring its status and trends. Compared with almost every other sector (trade, financial flows, health, climate), biodiversity is only monitored in the most rudimentary manner. As a result, we know little about relative rates of biodiversity loss around the world, and even less about the vulnerability of species and ecosystems to the various factors causing this loss. We can not assess the medium- and long-term effects of human activities on biodiversity. To fill this gap, the Species Survival Commission of IUCN-The World Conservation Union, in association with several other organizations, has been developing the *IUCN Red List of Threatened Species* as one of the tools to assess and monitor the status of Earth's biodiversity. As the world's most comprehensive, authoritative list of globally threatened species, the Red List not only provides information on the status of individual species and species groups; as it expands its scope in evaluating extinction risk in species using objective, quantitative criteria applicable across all species groups, including some that have never before been assessed, in particular marine species and many groups of invertebrates, it is becoming a global index of the status of biodiversity in general.

There can be little disagreement about what it will mean to lose forever the giant panda, any of the world's five rhinos, the Asian or African elephant, a lady's slipper orchid, or a great whale. But these species, conservation's "poster children," represent only a tiny minority of the world's endangered species. Although the conservation status of most of Earth's species has not yet been assessed, the current list —and the message it conveys of the extent to which we have already severely degraded the world in which we live— is overwhelming in its length and breadth. From islands to lakes, tropical rainforests and coral reefs, grasslands and rivers, animals and plants, vertebrates and invertebrates have disappeared —and are fast disappearing.

POPULATION EXTINCTION: A CRITICAL ISSUE

Few if any environmental problems are as important as the decay of biodiversity. A critical component of this is the loss of species. It is possible that in the current extinction episode, more than half of the world's plant and animal species will disappear in this century. This would be a catastrophe unparalleled in the history of humanity, and a disaster that might take ten to fifteen million years to recover from.

Abundant attention has been paid to the extinction of species, although little enough of the attention has been converted into solid policies to prevent it. In addition, as Dr. Jennifer Hughes and her colleagues have shown, a neglected facet of the extinction crisis is the rapid and accelerating loss of distinct populations of organisms. These components of species are disappearing much more rapidly than species themselves, and this erosion of population diversity has very serious consequences for the survival of species and for humanity directly.

Why should we be concerned about population extinctions? There are a number of reasons. The first involves ethics —the issue of whether human beings have the right to exterminate either species or components of species. In our view, and that of most concerned individuals, this is an important issue to be considered when any development or other human activity is going to cause extinctions. Of course, population extinctions are inevitable in connection with many human activities, be it building a new home in a previously undisturbed area or changing the global climate. But we should always be aware of those extinctions, try to limit them, and attempt to compensate for them by increasing population diversity elsewhere.

A second reason is the aesthetic and recreational value of populations. Père David's deer (*Elaphurus davidianus*) still exists as a species in a few reserves and zoos, but no human being has been able to enjoy observing them in nature for over a century. Similarly, the magnificent spectacle of 15 million bison roaming the great plains of North America is now lost even though *Bison bison* survives. Indeed, if Martha, the last surviving passenger pigeon (*Ectopistes migratorius*) were still alive in the Cincinnati Zoo, she would be a pathetic substitute for the many breeding colonies containing up to a billion individuals that once graced northeastern North America. If most animal species survived only in zoos, the joys of wildlife watching would be gone for everyone —it is populations that are critical to this hobby.

Perhaps the most important reason for human beings to preserve populations is their crucial role in providing humanity with ecosystem goods and services, as eloquently described by Dr. Gretchen Daily. Ecosystem goods are those products supplied to us by natural ecosystems, such as timber, fibers, game animals, fishes, medicines, pets, ornamental plants, spices, and so on. If a viable population of mahogany trees (*Swietenia macrophylla*) only existed in a botanical garden, that would be of little solace either to those employed by the timber industry who now make their living harvesting it, or to those who benefit from having its beautiful wood in their homes. From this perspective, it is not the existence of the species itself that is most important to us, but the existence of enough populations to support its commercial use over time. The same can be said for all the species of fish and other wild organisms that are harvested by humanity.

Ecosystem services are those vital services supplied free of charge to society by natural ecosystems. Without these services we could not survive. They include amelioration of the climate, running of the hydrological cycle, prevention of floods, generation and preservation of the soils that are critical to agriculture and forestry, pollination of crops, control of crop pests, and so on. Here again, populations are at least as important as species. If the population of *Pinus radiata* trees that grew on the hillside above your home was exterminated by a lumber company, it would be of little consolation when a flood or mudslide destroys your dwelling, that *Pinus radiata* still exists as a species. Ecosystem services are provided primarily by populations, and as species lose populations they also lose much of their value to humanity.

A prime example of the importance of populations is the role played by insects that pollinate crops. The existence of an effective pollinating species, by itself, tells you nothing of its value to agriculture. One must know the distribution and size of its populations before its contribution to our food supply can be estimated.

Another prime example of the need to conserve populations can be seen in the functioning of watersheds. For example, New York City recently decided that instead of spending six billion dollars on a filtration plant, it would restore the purity of its water supply by protecting and enhancing the natural ecosystem of its watershed. The latter was estimated to cost only 1.5 billion dollars. But of course, that enhancement meant saving many populations of trees and other organisms, the vast majority of which did not represent endangered species. Indeed, populations of the very same species could be involved in providing water purification services to many other watersheds. Population extinctions could have cost New York City many billions of dollars, and such extinctions are costing cities around the world amounts of at least that magnitude.

Obviously, another serious aspect of population extinctions is that they gradually contribute to species extinctions one bit at a time. The existence of many geographically scattered populations "spreads the risk" —for example, they make it much less likely that a catastrophic epidemic, fire, flood, volcanic eruption, or other major regional event would exterminate the species. This is often not considered when the effects of habitat destruction on species communities are measured. While destruc-

tion of, say, 90 percent of an area of habitat may still leave 50 percent of the species of the community surviving, that same destruction would usually wipe out 90 percent of the populations. That would leave many of the surviving species in a much more precarious position than before.

We should note that the drivers of species extinctions are also those of population extinctions. These, of course, are the multiplicative factors of the $I = PAT$ equation. Human Impact, in this case extinction, is a product of three factors: The size of the human Population, its per capita consumption (= Affluence), and the Technologies, including the economic, political, and social arrangements established to service that consumption. The $I = PAT$ equation means that any factors that reduce population pressures and limit wasteful consumption will help to protect our life-support systems. It also means that efforts to make technologies and social arrangements less environmentally destructive will have a similar effect.

Such efforts to decrease human impacts are absolutely critical if most biodiversity is to survive. But the payoff generally will not be seen for decades. Nevertheless, we must expand our efforts to preserve species diversity to give greater protection to population diversity. It is critically important both for the survival of Earth's great diversity of species and also for the survival of our civilization.

GERARDO CEBALLOS
Universidad Nacional Autónoma de México

PAUL R. EHRLICH
Stanford University

EXTINCTION: A NATURAL
—AND UNNATURAL— PROCESS

The world is, and always has been, in a state of flux. Even the land beneath our feet is constantly on the move. Over hundreds of millions of years, continents have broken apart, oceans appeared, mountains formed and worn inexorably away. These processes continue, barely discernible over a single human lifetime. With geological change come changes in living things: populations, species, and whole lineages disappear, and new ones emerge. The entire basis of organic evolution is underpinned by the appearance of some species and the disappearance of others; extinction is, thus, a natural process. According to the fossil record, no species has yet proved immortal; as few as 2%-4% of the species that have ever lived are believed to survive today. The remainder are extinct, the vast majority having vanished long before the arrival of humans.

The rapid loss of species that we are witnessing today is estimated by some experts to be between 1 000 and 10 000 times higher than the "background" or expected natural extinction rate. It is, thus, a most unnatural phenomenon. Worse, unlike the mass-extinction events of geological history, it is one for which a single species —ours— appears to be almost wholly responsible.

Extinctions in geological history

Although extinction is a natural process, it is a very variable one: the extinction rate has been far from constant through geological time. There have been long periods, of tens of millions of years or even longer, when few species died out, punctuated by phases of elevated extinction, when large numbers of species disappeared in a short, and sometimes very short, time. The most extreme of these periods are the mass-extinction events, five of which are recognized by paleontologists as having occurred in the 550-odd million years since animals first appeared in the fossil record. The most famous of these was the most recent, at the end of the Cretaceous some 66 million years ago, when perhaps as many as three quarters of the world's species, including all the representatives of three major groups of reptiles —the dinosaurs, marine plesiosaurs, and flying pterosaurs— vanished. However, the most serious took place nearly 200 million years earlier, at the end of the Permian, when over what was probably a relatively long period —perhaps 5-8 million years— higher life-forms are thought to have disappeared almost entirely: more than 95% of all species in existence at that time are believed to have become extinct, for reasons that remain a mystery.

to page 98 →

Extinctions and humans

Extinctions caused by humans are generally considered to be a recent, modern phenomenon. However, there is a compelling, albeit circumstantial, body of evidence suggesting humans started to play a significant part in the extinction of other species tens of thousands of years ago. One major indication is the unusually high extinction rates amongst large terrestrial animals, chiefly mammals and birds, but also some reptiles, in the past 100 000 years. These species, typically with a body mass of 50 kg or more, have been characterized as the "Pleistocene megafauna"; their demise in many parts of the world was very sudden and, crucially, is believed to have almost always coincided with the advent of humans.

The losses in some parts of the world were extensive. In Australia, where the earliest human remains are dated to *ca.* 64 000 years ago, no fewer than 19 of 22 identified genera of large land animals disappeared, the great majority between 30 000 and 60 000 years ago. Animals that vanished included giant wombats and kangaroos, an eight-meter-long monitor lizard (*Megalania*), and a huge, emu-like bird *Genyornis newtoni*. In the Americas, just under 80% of large-bodied genera became extinct. Extraordinary creatures, such as saber-toothed cats, mammoths, giant armored glyptodonts, and giant ground sloths, all disappeared some time between 11 000 and 13 000 years ago, coinciding with the dates of the first incontrovertible evidence of a human presence there.

The explanation for this extinction pattern has been hotly contested for decades between those who maintain that climatic changes were responsible and others who believe that humans were very largely, or entirely, to blame. In support of the former is the undoubted fact that global climates were very extreme, and very variable, during this period, which marked the end of the Pleistocene ice ages and the transition to the modern climatic conditions of the Holocene Epoch. In addition, in the case of the continental faunas, there is little direct evidence of major human impacts on these species. Countering this is the fact that these extinctions coincide with the arrival of humans in these parts of the world. This correlation is particularly strong in the case of the disappearance of island faunas in the Caribbean and New Zealand and on Madagascar. These regions had their own distinctive megafaunas —in New Zealand, in the form of giant birds known as moas; in Madagascar, as giant lemurs, dwarf hippopotamuses, and the extraordinary elephant bird *Aepyornis*; and on the larger Caribbean islands, as a series of large rodents and ground sloths. All these survived until much more recently than the continental faunas, and all seem to have disappeared within a few hundred years of the arrival of humans —in the case of the moas, within the last 300 or so years.

The precise mechanism —or mechanisms— by which humans may have caused these extinctions remains unclear. Overhunting, habitat destruction through extensive burning, and the introduction of new diseases (which are difficult to establish from this distance in time) have all been suggested. Extensive use of fire and direct hunting of animals that had evolved in the absence of humans and, thus, were not adapted to eluding their predatory behaviors, may have been sufficient to cause the extinction of the large herbivores such as the North American mammoths and the South American giant ground sloths. Populations of large carnivores and scavengers such as the giant, vulture-like *Teratornis* may then have collapsed as their prey-base disappeared.

Modern-day extinctions

However persuasive, or otherwise, the evidence might be, it is extremely unlikely that the real cause of the demise of the giants of the Americas and Australia in the closing millennia of the Pleistocene will ever be known. As we approach the modern era more closely, it becomes easier, though rarely easy, to record extinctions and understand their causes. It is a daunting task.

Cataloging extinctions depends on the use of historical evidence —written accounts and observations or actual remains of animals and plants— to reconstruct past events. This evidence is often scanty or inconclusive. Determining the taxonomic status of extinct populations on the basis of surviving evidence is often difficult: many apparently extinct species, particularly plants, may in fact have been simply populations of more widespread and still surviving species. Furthermore, demonstrating that a species is extinct is essentially trying to prove a negative. Time and again, species have been shown to be somewhat more tenacious than had been believed and have reappeared, sometimes after remarkably long absences. Notable examples include several species of bird, such as the Bermuda petrel or cahow *Pterodroma cahow*, believed extinct for three hundred years until rediscovered at the start of the twentieth century, and Jerdon's courser *Rhinoptilus bitorquatus*, last collected in India around 1900 and presumed extinct until found again in 1986, as well as a number of mammals, such as the Australian dibbler *Parantechinus apicalis*, unseen for 83 years between 1884 and 1967, and reptiles such as the Jamaican iguana *Cyclura collei*, thought to have gone extinct in the 1940s but rediscovered in 1990.

Unfortunately, these so-called "Lazarus species" are very definitely in the minority, and there is no doubt that the vast majority of the extinct species carefully cataloged by scientists over the past few decades have indeed gone for good. In the past five hundred years, an estimated 90 mammal species and over 100 bird species are known to have become extinct, along with over 200 other vertebrate and some 320 invertebrate species. Several hundred plant species are believed to have disappeared in the same period. Particularly striking is the fact that almost 250 of these vertebrate extinctions have occurred since 1900, and of these, almost 200 since as recently as 1950.

The overwhelming number of recorded extinctions in the past 500 years have taken place in two distinct areas: on islands and in inland water ecosystems (see Table 5).

Extinction on islands

The close of the Pleistocene did not mark the end of the megafaunal extinctions: the giant birds of New Zealand and Madagascar, as well as several outsized mammals on Madagascar, all survived well into the early modern era. The demise of the Malagasy species remains little-documented. In contrast, the extinction of the moas on New Zealand can be traced quite well, from the start of large-scale moa hunting around 900 years ago to its peak around 650 years ago and its petering out some 400 years ago, when there were virtually no more moas left to kill.

It has become increasingly clear in the past few years that it was not just the outlandish, oversized species like the moas that suffered at the hands of the early island settlers. Research on the Pacific Polynesian islands has revealed a range of bird species that existed at the time of the first Polynesian settlers, but which were extinct by the time Europeans began to colonize the region in the eighteenth and nineteenth centuries. On the Hawaiian Islands alone, at least 60 species of land birds, including flightless ducks, geese and ibises, birds of prey, and a host of finches, disappeared. On the basis of these and other figures, ornithologist David Steadman of the New York

State Museum has estimated that perhaps as many as 2 000 bird species may have disappeared since humans started to settle the Pacific in the last few thousand years.

Because of their light, fragile skeletons, birds often leave few, if any, identifiable remains when they die. Therefore, large numbers of species have undoubtedly disappeared without trace, and it will never be known for certain how many there were. However, those extinctions that have taken place in the past 400-500 years and are fairly well-documented have overwhelmingly been of island rather than continental species. Of the 200 or so known terrestrial vertebrate extinctions since *ca.* 1500, 85% were on islands: all but one of 20 reptiles, 63 of 85 mammals, and 97 of 107 birds. If Australia is regarded as a super-island, then this pattern becomes even more marked, as a further 18 documented extinctions (17 mammals and one bird) took place there, leaving fewer than twenty documented extinctions in the whole of Africa, Eurasia, and North and South America. Only amongst amphibians is the pattern different: all five recorded extinct species were continental forms.

One reason for this disproportion in the number of island extinctions may be the fact that it is somewhat easier to document extinctions on islands, particularly small ones, than in continental areas, where there is generally a greater likelihood that a species persists in a little-explored area, or occurs outside what had been thought to be its entire range. For this reason, a number of continental species that have not been seen for some time are regarded as probably or possibly extinct rather than categorically extinct, amongst them the pink-headed duck *Rhodonessa caryophyllacea*, last recorded in the wild in 1949, and the crested shellduck *Tadorna cristata*, last observed in the year 1964.

However, the high number of island extinctions is undoubtedly a real phenomenon and not simply a product of observational bias: over the last two thousand years at least, island species have clearly been far more susceptible to extinction than continental ones, for a number of reasons. First, island populations tend to be small and to have highly circumscribed ranges, with fixed boundaries limiting emigration, either natural or when under stress; it is, therefore, quite easy to exterminate them or deplete their populations to a level at which natural fluctuations in their numbers or catastrophic events can lead to extinction. Second, island species, particularly those on isolated oceanic islands, have usually evolved in the absence of ter-

to page *110* →

restrial predators and, thus, have generally lost any defensive mechanisms to ward them off. Plants appear to be as vulnerable as animals in this regard, as many island plants lack the deterrents to herbivores —such as thorns, toughened tissues, and noxious chemicals— that their continental counterparts possess. Island plants and animals alike are, therefore, often ill-equipped to tolerate nonnative species which act as predators or competitors, and have been introduced by humans in abundance. These introductions have been both accidental —such as rats, ants, and hosts of aquatic invertebrates and weeds— and intentional —including domestic crops and ornamental plants and domestic livestock and companion animals. Amongst the intentional introductions have been a number of experiments in biological control gone awry: mongooses, cane toads, and carnivorous snails were introduced in misguided attempts at "naturally" controlling various pests, many of which were themselves introduced, with disastrous consequences.

Beyond the introduction of other animals and plants have been a range of other impacts associated with colonization by humans, including hunting for food and destruction of habitat through logging and conversion to agriculture. In various ways, all these impacts have led to the demise of many distinctive and extraordinary island organisms over the past few hundred years. Perhaps the most famous of these is the dodo *Raphus cucullatus*, the giant flightless pigeon from Mauritius in the Indian Ocean, exterminated at the end of the seventeenth century by, it seems, a combination of wanton slaughter (the bird was neither a nuisance nor particularly good eating) and the effects of introduced pigs, cats, rats, and dogs. The dodo was one of three similar species in the region destined to meet the same fate: the Réunion solitaire *Raphus solitarius* died out in *ca.* 1710, while the Rodrigues solitaire *Pezophaps solitaria* survived for another half century.

One startling finding of the last few decades has been the devastating impact that a single introduced species can have on an island's biota. The brown tree snake *Boiga irregularis*, a native of the New Guinea region and northeastern Australia, was accidentally introduced to the Pacific island of Guam in the late 1940s or early 1950s. Within four decades, it had spread throughout the island, reaching a population density of up to 5 000 individuals per square kilometer. The native vertebrate fauna had no defenses against this aggressive arboreal predator, with the result that nine of the eleven native forest bird species, two of the three native bats, and five of lizard have now

been exterminated. Fortunately, most of these species survive on nearby islands which are as yet free of the snake, but three —the Guam flying fox *Pteropus tokudae*, Guam rail *Rallus owstoni*, and Guam flycatcher *Myiagra freycineti*— were endemic to that country. Only the second one survives in captivity; the other two are gone forever.

On Moorea in the Society Islands of French Polynesia, similar damage was done, albeit unwittingly, by a deliberate introduction. In 1977, a carnivorous snail from Florida, *Euglandina rosea*, was introduced in an attempt to control the giant African land snail *Achatina fulica*, itself an introduced species and a serious crop pest. Unfortunately, the carnivorous snail preferred to feed on the smaller, more vulnerable native snails in the genus *Partula*, part of a unique Pacific radiation of land molluscs representing an ancient evolutionary lineage. Within twenty years, all seven native *Partula* snails were extinct, while the giant African land snail continued to thrive. This pattern has been repeated elsewhere: the *2000 IUCN Red List* includes no fewer than 57 *Partula* and over 150 other land snails, almost all island forms, as Extinct. *Euglandina* appears to be responsible for the demise of many of these, but others have fallen victim to a range of introduced predators including fire ants and flatworms. A mere handful of the species (including five Moorean *Partula*) survive in captivity as the result of last-ditch captive-breeding efforts.

It is not just animals that have suffered extinctions on islands. So devastated are many oceanic islands that little is known of the nature of the original vegetation. However, where it has been possible to reconstruct a picture of the original flora, it is often evident that many species have been lost. The island of Saint Helena in the South Atlantic Ocean is believed originally to have supported some 100 endemic plant species. Ten years after the island was discovered in 1502, a small herd of goats was introduced, which rapidly turned into a population of several thousands. The effects of these, combined with tree felling for construction and fuel wood, meant that by 1810 the original dense forest cover was reduced to a few fragments on inaccessible ridges. Of the original endemic plants, fewer than forty now remain, and half of these are seriously threatened with extinction.

Terrestrial continental faunas

While the recorded rate of vertebrate extinctions in the last 500 years has been far higher on islands than on continents, these latter extinctions are also of great significance. In Australia, European colonization appears to have initiated another wave of mammalian extinctions following the demise of the megafauna thirty or more thousand years before. Although habitat alteration clearly played a major role, the coup de grâce was apparently delivered in most cases by introduced cats and foxes. In addition to seventeen extinct mammals, several others, such as the western barred bandicoot *Perameles bougainville* and the banded hare wallaby *Lagostrophus fasciatus*, are now confined to predator-free offshore islands, while yet others, including the numbat *Myrmecobius fasciatus*, survive on the mainland only through intensive predator-control programs.

In North America, the most famous recent extinction is undoubtedly that of the passenger pigeon *Ectopistes migratorius*, a colonial nesting bird of extraordinary abundance whose population collapsed during the nineteenth century due largely, it seems, to relentless hunting. The last-known individual, a captive-bred female named Martha, died in 1914 in the Cincinnati Zoo at the age of 29. In Africa, two antelopes from opposite ends of the continent disappeared over 100 years ago: in South Africa, the bluebuck *Hippotragus leucophaeus* was exterminated in *ca.* 1800, almost certainly a victim of overhunting, while in Algeria, the little-known red gazelle *Gazella rufina* seems to have vanished towards the end of the nineteenth century. In Europe, the Bavarian pine vole *Microtus bavaricus* was last recorded in the early 1960s, when a hospital was built over its remaining habitat.

To the twenty-odd extinct continental species must be added a number that have disappeared from the wild but survive in captivity or have been the subject of reintroduction programs, such as the scimitar-horned oryx *Oryx dammah* of North Africa and its close relative the Arabian oryx *Oryx leucoryx*.

Extinction in inland waters

From the point of view of the species that inhabit them, inland water ecosystems behave rather like islands, in that they are typically areas of limited extent isolated from other such areas. This can be most clearly seen in desert springs, but also in lakes and the catchment areas that feed them. Populations of wholly aquatic species in these water bodies are usually cut off from any other populations and have a strictly circumscribed range. River systems, at least those that run into the sea, are somewhat different in that aquatic species which tolerate salt water have an opportunity to disperse through the sea into other rivers; however, for the substantial number of species that are intolerant of salt water, the sea at the river's mouth serves as a barrier as effective as the land.

Just as island species are particularly vulnerable to external influences, so are these aquatic forms. Draining of lakes, diversion and channelization of rivers, impoundments and other physical barriers, pollution, and injudicious introductions can wipe out a species or destroy its entire habitat in a very short period. Because our knowledge of freshwater species is very incomplete and monitoring their numbers and distribution is so much more difficult than on land, it is probable that the number of extinctions from these ecosystems has been and continues to be severely underrecorded. There is evidence to suggest that these species are suffering one of the highest rates of species loss. In North America, where freshwater biodiversity is not only particularly rich but has been well studied for over one hundred years, at least 123 freshwater animal species have gone extinct since 1900, and hundreds of others have been documented to be at risk of extinction. A recent study derived current and future extinction rates for North American freshwater fauna that are at least five times higher than those for terrestrial animals.

The most dramatic recorded collapse of any vertebrate fauna in the past hundred years has undoubtedly been that of the fishes of Lake Victoria, East Africa. This huge lake supported until recently an extraordinary radiation of several hundred species of closely related cichlid fishes, many of them beautifully-colored and including a range of peculiar, highly-specialized forms. This fauna was of immense scientific interest because of the potential insights it gave into the mechanisms of evolution. However, in the late 1950s, the large, predatory Nile perch *Lates niloticus* was introduced to boost fisheries production in the lake. So successful was the perch that within twenty-five years, it had spread throughout the lake and caused the complete disappearance of over one hundred of the endemic cichlids. Populations of the remaining species shrink yearly in the face of predation by this voracious interloper and a host of other threats including destructive fishing methods, sedimentation, and the spread of the introduced South American water hyacinth *Eichhornia crassipes*, a pernicious weed that is suffocating water bodies throughout the tropics.

From what we know, it seems that the catastrophe of Lake Victoria is being repeated, albeit at a lesser scale, in many other lakes

to page *118* →

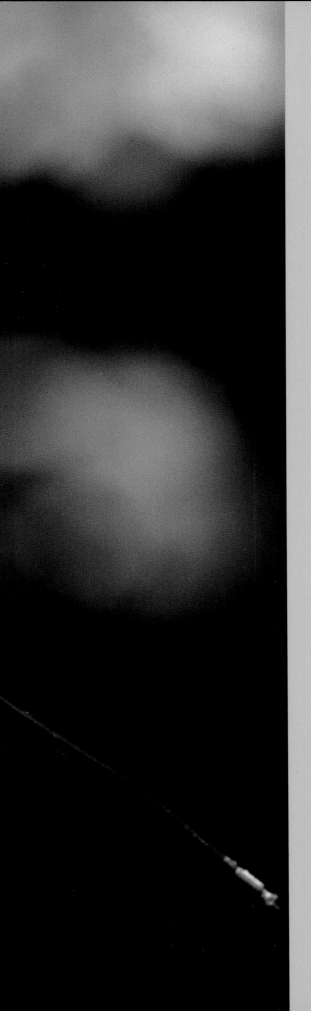

and river systems around the world. In Lake Lanao on Mindanao Island in the Philippines, at least fourteen species of endemic fish disappeared some time before 1982 while in Mexico, at least a dozen fishes, most of them confined to isolated desert springs, have been lost in the past few decades. The loss of freshwater invertebrates, such as the North American unionid mussels, is as or more severe.

Recent extinction in the seas

The problems of proving extinctions on land and in fresh waters are much more challenging for marine ecosystems, which are less well known and studied, and remote. Moreover, it has long been thought that marine species were immune to extinction, owing to a number of characteristics —real and presumed— that would ensure that they could well withstand heavy fishing and other threats. Until recently, the list of extinct marine species was limited to a few large vertebrate species closely associated with inshore shallow waters, namely Steller's sea cow *Hydrodamalis gigas*, or dependent on land to breed, such as the Caribbean monk seal *Monachus tropicalis* and the great auk *Pinguinus impennis*. No truly marine species had yet been proven to have disappeared in modern times. However, this situation is now changing as more species are added to the extinction list. The first well-documented case of a marine invertebrate's extinction is that of the eelgrass limpet *Lottia aelvus*, which disappeared in the 1930s from the northeast coast of North America when an epidemic wiped out the eelgrass beds that it occupied. Two other North American marine invertebrates —another limpet *Collisella edmitchelli* and a horn snail *Cerithidea fuscata*— are also considered extinct, while a third, the white abalone *Haliotis sorenseni*, is on the verge of extinction, having been reduced to a few tens of individuals from a population that numbered 2 000-10 000 individuals per hectare in the 1970s. The survival of several restricted-range coral-reef fishes has recently been questioned by scientists, suggesting that they may soon be added to this list, as may a number of Critically Endangered fishes featured in the IUCN Red List, such as the speckled hind *Epinephelus drummondhayi* and the large-toothed sawfish *Pristis perotteti*. These examples and a growing body of evidence on the vulnerability of countless marine species leave little doubt that the number of documented marine extinctions will grow.

The major threats to species

Virtually all the factors that have led to the extinction of species in the modern era evidently continue to operate, many with ever-increasing intensity. While species on islands and in fresh waters still appear to be the most highly threatened, it is evident that the number of species under threat on continents and in the seas is growing rapidly as human activities extend their impact on even widespread and previously abundant organisms. The current threats to species are manifold, with different threats often interacting and reinforcing each other. While diverse factors vary in intensity and relative importance in the three major biomes (the land, inland waters, and the seas), certain common threads emerge.

Habitat loss and degradation

The most pervasive and overriding threat to terrestrial, freshwater, and coastal marine species is indisputably the loss and degradation of habitats through human activity. Conversion of land to agriculture; the spread of human settlements; industry; infrastructure such as roads, dams, airports, and port facilities; and extractive activities such as mining, logging, and fisheries are directly affecting an increasing proportion of the Earth, leaving fewer natural habitats and ecosystems intact and destroying others. The diversion of water for irrigation, for example, has caused the demise of the Aral Sea in central Asia; once the world's fourth largest inland body of water, it is now drying and barren, and predicted to disappear altogether in ten years' time. Aside from the direct impact caused by conversion, there are indirect impacts such as the effects of pesticides and other chemicals, noise and other disturbance, and effluents, either organic or inorganic. With the human population projected to rise from six billion to nine billion during the coming century, these impacts can only increase.

Effects of alien invasive species

Although generally less obvious a factor in the loss of biodiversity, the effects of alien invasive species may be second only to habitat loss in terms of the extent and severity of their impact. They have played a pivotal role in the extinction of many species, and all the evidence indicates that they are becoming ever more pervasive as increased trade and travel around the world breaks down the physical barriers that have naturally separated species and facilitates their movement. In many instances, introduced species —freed from the predators and other factors that limited their spread in their natural habitats— proliferate at a far greater rate than they would naturally. These invasive species are not only plants and animals but also pathogens, and their spread is having disastrous effects on species and ecosystems —and economies. Not only are the effects of alien invasive species jeopardizing a growing number of species and ecosystems, they are threatening to homogenize the planet.

While the impacts to date in terms of actual extinctions have been largely on islands and in fresh waters, alien invasive species are becoming an increasing problem on continents and in marine ecosystems. For example, the roster of species that have invaded marine areas as a result of human activity is already extensive and is growing daily. It is estimated that some 400-500 species have migrated through the Suez Canal from the Red Sea into the Mediterranean Basin since the canal was opened in 1869, with new species believed to arrive in the Mediterranean at the rate of 4-5 annually. In the Black Sea, one species —Leidy's comb jelly *Mnemiopsis leidyi*, introduced from its native waters on the Atlantic coast of the Americas in ships' ballast water— has contributed to the collapse of the entire Black Sea ecosystem, including its fisheries. A Pacific seaweed, *Caulerpa taxifolia*, has carpeted over 15 000 hectares of the seafloor of the Mediterranean and Adriatic, doing untold damage to rocky inshore habitats in the fifteen years since it accidentally escaped from the Oceanographic Museum in Monaco.

Overexploitation

Humans use wild species for an almost endless array of purposes ranging from the fundamentals of food, clothing, shelter, and medicines to such "nonessentials" as cosmetics, ornament, and amusement. These uses can be local and subsistence, or global and highly commercial. Depending on the species and the nature and extent of other factors affecting it, any of these uses can represent a threat to species' survival. While management programs exist for many species, most of the world's wild plant and animal species are taken from the wilderness in the absence of any controls or monitoring to ensure that this exploitation does not deplete species' numbers. For many species, the effects of exploitation are unknown, but are suspected to be a problem. Judging to what extent and how to address it, with limited knowledge of species' status and ecology and scant resources,

presents a serious challenge to wildlife managers, in particular as, in most cases, exploited species are subject to other pressures such as habitat loss.

It is increasingly clear that local uses can deplete wild resources and that these uses can often be heavily commercial. This is the case with bushmeat consumption in the world's tropics, which affects a wide range of species from primates to wild pigs and peccaries, pangolins, hornbills, parrots, and pheasants, as well as many reptiles and amphibians. In West Africa, for example, hunting for local consumption and commercial markets has become the most immediate threat to apes and other primates and antelopes in the Congo Basin and has caused widespread local extinctions throughout the region. An additional pressure comes from international markets, where the demand may far exceed the supply. International trade patterns can change rapidly, shifting from one group of species or one country to another, often with little opportunity for safeguards to be put in place. Hence, controls on the trade in parrots and other wild birds, for example, appear to have prompted an increase in trade in reptiles, many of them little-known and localized species. The expansion in East Asian economies over the past few decades led to an increased trade in a number of wildlife commodities that has affected species around the world. The demand for products to be used in traditional Chinese medicine prompted a rise in poaching of threatened species such as the tiger *Panthera tigris* and black rhino *Diceros bicornis*, and also led to increased trade that has threatened numerous other species. The trade in Asian chelonians (turtles, tortoises, and terrapins) for food and traditional Chinese medicine, for example, is considered the major threat to the region's chelonian fauna.

While the trade in tropical timber has long been a rallying point for conservationists, overexploitation in targeted fisheries is now a concern, as it is a major threat for many marine species, including the humphead wrasse *Cheilinus undulatus*, captured for the trade in live reef fishes for food in East Asia; sea horses (family Syngnathidae), traded for traditional Chinese medicine, the aquarium trade, and for curios; and many of the world's elasmobranch fishes (sharks, skates, and rays). It is particularly severe, however, when combined with incidental take, or bycatch, in fisheries. Direct and indirect exploitation through bycatch in fisheries is considered to be a major problem for the world's cetaceans (whales, dolphins, and porpoises), including the Endangered Hector's dolphin *Cephalorhynchus hectori*, and marine tur-

to page *128* →

tles, including the Critically Endangered leatherback turtle *Dermochelys coriacea*, as well as many elasmobranchs. Bycatch is also a problem for species not targeted in fisheries, such as pelagic seabirds, suffering mortality in longline fisheries, or no longer targeted, such as the barn-door skate *Dipturus laevis* and common sawfish *Pristis pristis*, which continue to be taken in fisheries aimed at other species.

Conflicts with humans

The growing pressure for living space and on resources for food brings humans and wild species increasingly into conflict, particularly in the case of predators and larger animals. African elephants *Loxodonta africana* raid crops and may even destroy homes and kill people, while wolves *Canis lupus*, large felids —including cheetahs *Acinonyx jubatus* and leopards *Panthera pardus*—, and crocodilians prey on livestock and, in some instances, on people. Otters, seals, dolphins, and cormorants are accused of depleting fisheries resources, predating on fishponds and fish catches, and damaging fishing gear. Persecution by humans is, thus, an increasing problem for numerous species, including many that are at risk of extinction. Resolving these conflicts, whether they are real or merely perceived, is a major challenge for wildlife managers and conservationists today.

Pollutants

While the impacts of habitat destruction or overexploitation are relatively easily understood and monitored, it is much more difficult to assess the effects of pollutants. They are often invisible, can disperse widely from their point of origin, and their effects frequently accumulate over time, so that they may not become apparent until it is too late to redress those effects.

Major pollutants include nutrients such as nitrates, nitrites, and phosphates, and organic matter such as sewage; persistent organic pollutants (POPs), including a range of chlorinated hydrocarbons; and heavy metals such as cadmium, mercury, and lead. Many heavy metals and POPs can act as toxins above certain concentrations, causing mortality or reducing reproductive success, particularly in cases where they become increasingly concentrated towards the top of food chains, as has been found with marine mammals, certain fishes, and other predators. The best known example of this is undoubtedly the effect of organochlorine insecticides, most famously DDT but also a variety of others that became very widely used in the

1950s. These accumulated in the food chain and built up to very high levels in top predators, such as falcons and other birds of prey. While some individual birds undoubtedly died as a direct result of this accumulation, the major effect was to cause females to lay eggs with thinner, fragile shells. The eggs almost invariably cracked before incubation was over, causing almost complete breeding failure in many bird populations in Europe and North America. Action has been taken to control the use of these insecticides in some parts of the world, but these and many other chemicals whose long-term impacts are unknown continue to build up in the biosphere.

In aquatic environments, buildup of organic nutrients can have a catastrophic effect, leading to algal blooms that may be toxic to other organisms or that use up available oxygen, resulting in mass die-offs of most animal species. Zones with little or no dissolved oxygen in the water are increasing in size and frequency in many estuarine areas around the world, with disastrous consequences for fishes and marine invertebrates, and the fisheries that depend on them.

One major class of pollutants whose effects are worldwide are greenhouse gases, chiefly carbon dioxide, methane, and nitrous oxide. The effects of these gases have been the subject of great controversy. However, a consensus has emerged in the world's scientific community that buildup of these in the atmosphere as a result of human activities will have a significant effect on the world's climate in the next century and beyond. We can only begin to speculate what the impact of this will be on natural ecosystems and the biodiversity they support, but there are already indications that much of this impact will be negative. Likely to be particularly badly affected are polar ecosystems and high mountain areas. Already, desiccation of cloud forests as a result of changes in the cloud base has been implicated in the decline or even disappearance of some amphibian species such as the golden toad *Bufo periglenes*, while warmer temperatures in the Arctic affecting the polar ice cap have been found to be negatively influencing prey availability and, hence, the physical condition and reproduction of polar bears *Ursus maritimus* in at least one population in Canada.

Disease

Disease is a natural phenomenon and, under normal circumstances, one that is unlikely to lead to the extinction of a species. This is because diseases generally become self-regulating: if the disease-causing organism becomes too virulent and destroys its host too quickly, it itself will probably die before it has a chance of being transmitted to another individual. In addition, the selection pressure on organisms faced with disease is so great that at least partial resistance is likely to spread through populations rapidly. However, where populations of species have already been reduced to critically low levels by other causes, disease may become an important factor in determining their survival. The formerly common white-rumped vulture *Gyps bengalensis* of the Indian subcontinent is now Critically Endangered in part because of the effects of viral disease, while various diseases have been identified as major agents in mass mortalities of amphibians around the world. An apparent increase over the past 30 years in disease outbreaks caused by new pathogens in populations of wild species as well as humans is the subject of intensive study and extensive discussion in the scientific community, with certain scientists suggesting that this could be indicative of some major ecological change that has far-reaching implications. The authors of one recent study on the subject have noted that "pathogen pollution" is becoming an increasingly serious threat to biodiversity.

Averting future extinctions

Little can now be done for the species whose extinction we have already caused, and it is arguable how much can be done for most of the 2 000 or so Critically Endangered species whose extinction appears imminent. Yet every surviving population, no matter how small or imperiled, offers at least some hope for the future. In the face of the sheer number of species deserving our attention, and the multiplicity of threats they confront, it would be easy to become overwhelmed. It is vital, therefore, that conservation action be planned using all the information at our disposal and, in particular, that we have as accurate as possible an assessment of the extinction risk faced by any given species. This is what the IUCN Red List sets out to provide.

EXTINCTION WAVE IN THE MAKING

The freshwater bivalves, or pearl mussels, that make up the order Unionoida are the world's most endangered group of animals. In the southeastern United States, where the greatest pearl mussel diversity —269 species— occurs, no fewer than 34 species have gone extinct, and dozens more are threatened with extinction. The remainder face a doubtful future. Their specific case, a "mini" extinction wave in the making, is not an isolated event: other freshwater species, subject to the same threats in North America and elsewhere, are also disappearing before our eyes.

Some 1 000 pearl mussel species are found in freshwater lakes and rivers on six of the world's seven continents. They are unique and remarkable in a number of ways. Their reproductive cycle includes a period during which their larvae, called glochidia, "hitchhike" on the gills and fins of fish until they are old enough to survive on their own. More often than not, one species of mussel parasitizes in this way a single species of fish; one species even uses the North American mud puppy, an aquatic salamander, for this purpose. Freshwater pearl mussels are also among the longest-living animals: they become sexually mature at about six to seven years of age and live for decades; some species, like the European pearl mussel, have been known to live as long as 200 years.

Freshwater mussels were first reported to be declining in North America during the mid-nineteenth century, but it was not until the 1970s that any of these species were recognized as having become extinct. At that time, biologists noticed about 11 species that had not been collected in the past 50 years. Today, some 35 of the 300 species in North America are presumed to have become extinct in the last 70 to 100 years. Many are declining and some are functionally extinct, for the living animals are senescent and no longer reproducing and at least one species is down to a single population. Due to a lack of complete field surveys, we do not know for sure how many, but most certainly, some species are on the brink of extinction.

The decline, decimation, and final extinction of the freshwater pearl mussels of North America are a direct result of the destruction and degradation of the rivers and streams that they inhabit. Damming of rivers, dredging, and pollution, including domestic, industrial, and acid runoff from coal mines, are just some of the many insults to which these ecosystems have been and continue to be subjected. In some areas of the eastern United States, streams that had a diverse freshwater mussel fauna historically have been so severely impacted by acid mine drainage that they are now home to a few aquatic flies and nothing more. The damming of rivers affects mussels in several ways. Most of these mussel species evolved in shallow riffle or run habitats with clean and highly-oxygenated water. Damming of the river impounds the water; dissolved oxygen drops to zero; sedimentation from the exposed shoreline may cover the mussels and smother them; the cold water temperature keeps them from reproducing; and just as importantly, the modified ecology causes the fish species that serve as hosts for the glochidia to leave the area. Without a fish host available, the reproductive cycle is broken: the mussel lives out its life, and the population is extirpated when the last individual dies. In other instances, damming or other impacts cause the extirpation of the host fish, with the same effect —the subsequent extirpation of populations and possible extinction of the freshwater mussel species.

Although the destruction of their freshwater habitats continues, the heaviest blow to North America's pearl mussels may well be from invasive species. The first invader was the Asian clam, which was initially introduced as a food item but quickly spread throughout North America, where it is believed to outcompete the pearl mussels for living space and food resources. The next two interlopers, the zebra mussel and quagga mussel, which originated from the Black and Caspian seas of Eastern Europe, have proven to be even more of a problem. Since 1986, when the zebra mussel was first reported, it has spread throughout the Mississippi, St. Lawrence, and Hudson River basins, reaching enormous densities. It is credited with the extirpation of the freshwater mussel fauna from Lake Erie and considered a major threat to the continued existence of many mussel species in the Mississippi River Basin. A preferred substrate for the zebra mussel is the exposed posterior end of a native freshwater mussel. As many as 11 000 zebra mussels have been found encrusted to a single pearl mussel, far more than is needed to overwhelm it and cause its demise.

Why should we care about the extinction of the yellow-blossom, the sugarspoon, the orangefoot pimpleback, the Cumberland monkeyface or the golfstick pearly mussel? Most people have a hard time developing any empathy for a clam that spends most of its life at the bottom of a river, however fascinating its sex life may be. Yet these sedentary animals provided the shell that served as the basis of the pearl-button industry which flourished in the American Midwest in the first part of the twentieth century. And these same shells are used as the seed material for the world's cultured pearl industry today. As important is their role in purifying the water of the rivers and lakes that we expect to drink from and swim and fish in. Freshwater mussels filter 1.7-2.5 liters of water per hour and are often found in populations of 200-400 individuals per square meter. Over the course of a few kilometers of river, this represents a significant filtration and cleaning of the water. Finally, pearl mussels fulfill a number of other ecological roles, including as food items for fish, turtles, muskrats, raccoons, and river otters; as hosts for other invertebrates; and, in at least two instances, brooding sites for developing fish larvae. Hence, their loss has a range of implications, ecological and economic, local and global.

The freshwater gastropods, or snails, of the southeastern United States —periwinkles or hornsnails and ramshorn snails— were also extremely diverse historically but have experienced a similar wave of extinction. About 42 species of North American snails are presumed extinct, including four entire genera, all confined to the Coosa River Basin in eastern Alabama. Much less is known of freshwater molluscs elsewhere in the world, but the destruction and degradation of freshwater systems from the same threats as in the United States give cause for serious concern that these species are disappearing just as rapidly and inexorably. And there is every reason to believe that the implications are no less important than with the freshwater pearl mussels of North America.

Recent efforts in the United States to expand riparian buffer zones, control mine runoff, close point sources of pollution, and otherwise improve water quality are enabling the return of fish and freshwater mussels to formerly polluted waters. However, the return of these long-lived mussels is a slow process and may take as long as 100 years, making it impossible to predict whether more species will be lost than preserved and, thus, how devastating a tragedy this extinction wave will be.

ARTHUR E. BOGAN
IUCN/SSC Mollusc Specialist Group

THE RED LIST
OF THREATENED SPECIES: DOCUMENTING
THE EXTINCTION CRISIS

Just as our knowledge of extinct species is very incomplete, the conservation status —or degree of risk of extinction— of most of the world's species is poorly known. Only a very small proportion of Earth's described species have been evaluated for their extinction risk, and most of these are terrestrial vertebrates and plants or species found in the better-studied and less remote parts of the world. The IUCN Red List is the world's most comprehensive compilation of species at risk of extinction, but for most species groups it is seriously incomplete. Nevertheless, it provides a compelling body of evidence on the imminent —and nearly imminent— loss of thousands of species. There is no more convincing document attesting to the current extinction crisis.

The most recent Red List, the *2000 IUCN Red List of Threatened Species*, classified just over 11 000 species as threatened with extinction; 816 as Extinct or Extinct in the Wild (see Table 6); and 4 600 as Data Deficient or Lower Risk: near threatened or conservation dependent. Although the list included several thousand threatened plants, mostly trees, the *1997 IUCN Red List of Threatened Plants*, still the working volume for plants, is more complete, listing 34 000 plants or 12.5% of the world total as threatened. Although the 11 000 threatened species featured in the 2000 Red List represent less than 1% of the world's described species, they include 24% of all mammals and 12% of all birds. For other vertebrate groups, the proportion of threatened species is believed to be as high or even higher; extrapolation from the species that have been assessed in these other groups suggests that as many as 25% of reptiles, 20% of amphibians, and 30% of freshwater fishes may be threatened with extinction.

The Red List: A global effort
to document species at risk of extinction

Conservation efforts aimed at individual species have a long history, but it was not until the second half of the twentieth century that serious attempts were begun to identify species at risk of extinction on a global basis. Pioneering work was initiated in the late 1950s by the (then) Survival Service Commission (now the Species Survival Commission) or SSC of IUCN–The World Conservation Union to develop a system for classifying species on the basis of how highly threatened they were considered to be and, simultaneously, to compile an inventory of threatened species worldwide. The aim was to provide guidance as to species that were at risk of extinction and the relative urgency of action required to avoid their loss. In the 1960s,

to page *148* →

this effort led to the publication by IUCN of the first in a series of Red (for danger) Data Books, cataloging species at risk of extinction and classifying them according to the relative severity of the extinction risk. In 1986, the first IUCN Red List of Threatened Animals was published; in contrast to the Red Data Books, which provided detailed accounts of the biology and status of each species, the Red List aimed to serve as a comprehensive global inventory of threatened species and did not include detailed species information.

From these early, tentative beginnings, the Red List has grown. The IUCN Red Data Books and Red Lists are now universally recognized as the global authoritative guides to the status of threatened species, as are those produced by SSC's partner organization on the Red List, BirdLife International, the world authority on the status of threatened birds. In addition, the Red List concept has been widely applied at regional, national, and local levels in both scientific and policy contexts. The system has evolved considerably since its inception, not least in the formulation and use of the threatened species categories, which were such an important part of the original concept.

Assessing extinction risk: The IUCN Red List System

The categories of threat for classifying species in the initial IUCN Red List system were: **Endangered** for the most highly threatened; **Vulnerable** for somewhat less threatened species that were thought likely to become endangered in the near future unless the factors affecting them changed; **Rare** for those thought at risk by virtue of their small populations or small ranges, or both; **Indeterminate** for those believed threatened but for which there was not enough information to assign them to one of the other categories; and **Insufficiently Known** for those suspected to be threatened, but for which there was insufficient information to make an accurate assessment. In addition, the system included three other categories: **Extinct**, **Not Threatened**, and **Out of Danger**, the last of these intended to apply to those species that were formerly classified as threatened, but whose status had now improved as a result of conservation action so that they were currently regarded as "safe."

This assessment system served its purpose very well for nearly thirty years, but over time, it became evident that changes were needed. The vagueness of the definitions of the categories and the lack of clear criteria for assigning species to them meant that they could be interpreted in different ways by different people. There was,

thus, no overall standard that enabled comparison between species or species groups or an assessment of how their status had changed over time. Further, advances in understanding of the extinction process suggested that a more quantitative system could be developed that would not only address these problems, but place the assessment process on much more solid scientific grounds in terms of estimating actual extinction risk. Led by population biologist Dr. Georgina Mace of the Institute of Zoology in the U.K., the Species Survival Commission developed a new set of Red List Categories and Criteria that were adopted by IUCN in 1994. The first application of this new system resulted in the *1996 IUCN Red List of Threatened Animals* and marked the turning point for the Red List. The new system has greatly enhanced the effectiveness of the Red List not only in assessing extinction risk for individual species, but also as an indicator of broader trends in the status of species and biodiversity. A number of revisions in the system, adopted by IUCN in 2000 and entering into use in 2001, mark a further advancement.

The 1994 Red List system was developed from the perspective of population biology —it assesses the probability that a species will become extinct some time in the relatively near future on the basis of various attributes of its biology and the status of its populations. Three major attributes are used: the area the species occupies (and how fragmented that area is); the size of its population; and the rate of change of the population. Each of these is perfectly intuitive: the smaller (or more fragmented) the area a species occurs in, the more susceptible it is to wholesale changes in its habitat, or to a single catastrophic event such as an oil spill or the introduction of a disease or a new predator. Similarly, the smaller a population is, the more likely are random fluctuations —which affect all species, common or rare— to drive it to a level at which it can no longer successfully reproduce. For many animal species, problems of inbreeding are also liable to manifest themselves when populations become very small; when this occurs, inherited abnormalities or weaknesses, such as lack of resistance to a particular disease, may spread throughout the entire population, leaving it permanently at risk even if numbers increase again. Finally, even an apparently widespread and abundant species may be seriously threatened if its population is declining rapidly enough —the classic example of this being the passenger pigeon, which plummeted from untold millions to zero in a few short decades.

On the basis of one or more of these attributes, species are assessed and placed in one of several categories, reflecting the estimated severity of the extinction risk. The three Categories of Threat in declining order are: **Critically Endangered**, **Endangered**, and **Vulnerable**. In addition to **Extinct** and **Extinct in the Wild** are the categories **Lower Risk** (which itself includes three subcategories of **Conservation Dependent**, **Near Threatened**, and **Least Concern**) and, importantly, **Data Deficient**, referring to species for which there is a lack of adequate information to make an assessment. The Lower Risk subcategories of Conservation Dependent and Near Threatened, and the category Data Deficient, although not categories of "threat," are very important: the two Lower Risk categories serve as a sentinel, calling our attention to species that could become threatened unless action is taken —or, in the case of Conservation Dependent, maintained— to prevent that from happening, while Data Deficient species are those for which new information may urgently be needed: these species could be threatened —even critically so— or not threatened at all.

The Red List system sets out quite precise criteria for inclusion of species in each of the three threatened species categories. To qualify for listing as Critically Endangered, for example, on the basis of a population decline, a species must be reliably believed to have declined by at least 80% over the last ten years or three generations

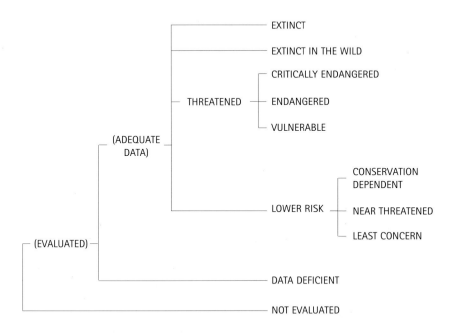

Figure 1. Structure of the categories.

(whichever is longer). Similarly, to qualify on the basis of population size, it must have a population estimated at fewer than 250 "mature individuals" and meet one of a set of other criteria, for example, a continuing decline of at least 25% within three years or one generation (whichever is longer). Extremely rare species —those whose population is estimated to number fewer than 50 individuals— automatically qualify as Critically Endangered.

The Red List: The most comprehensive assessment of globally-threatened species

For a handful of species that are the subject of intense conservation efforts, literally every individual may be known. These species, such as the Chatham Island robin *Petroica traversi* of New Zealand and the North American whooping crane *Grus americana*, typically have entire populations of no more than a few dozen individuals. Their status is unequivocally clear, and is often tracked on a continuous basis; hence, there is little difficulty in assessing their extinction risk and assigning them to an appropriate Red List category. However, no such certainty applies to the overwhelming majority of the world's species. Monitoring the status of wild populations of animals and plants is a difficult and time-consuming task, which is carried out by a relatively limited number of people and conservation agencies. Only a small fraction of the species that are of concern are actually monitored in any way. For the remainder, assessment of their status must be made by inference —often on the basis of a declining area of habitat, an observed population decline in one part of the range or, in the case of heavily-traded species, an increasing scarcity in markets. This often means that the classification is provisional, pending new data; however, one of the improvements in the Red List system is that the basis for listing is explicit, so that reclassifications in such instances can be done in a straightforward manner.

Given these constraints, it is not surprising that the conservation status of only a minute proportion of the world's species has yet been assessed. It is equally unsurprising that the proportion of species evaluated in different taxonomic groups varies enormously. Amongst animals, only the mammals and birds have been comprehensively assessed. The proportion of species assessed in other vertebrate groups is very low: fewer than one in six amphibians and reptiles, and one in ten fishes. The situation with invertebrates is even worse: fewer than one in twenty known species of mollusc and crustacean, and only one in every thousand of the described species of insect,

to page *158* →

have been assessed. In plants, just over 10 000 species, or only some 4% of the world's total, have been assessed utilizing the new IUCN categories.

Even with the severe limitations of our knowledge, the body of evidence presented by the Red List is striking (see Tables 7 to 11).

Birds and mammals

Because they are so much better known than other species groups, the status of birds and mammals provides a useful guide to the status of animal species in terrestrial environments, where the great majority of both birds and mammals occur. According to the *2000 IUCN Red List*, over 1 100 mammals (nearly one in four) and about the same number of birds (around one in eight) are threatened with extinction. Of these, almost exactly the same number of each (180 mammals and 182 birds) are classified as Critically Endangered. In both groups, there is a marked tendency for small, distinctive families and orders to have more threatened species than expected. In the mammals, for example, this applies to the sirenians (the dugong *Dugong dugon* and the manatees *Trichechus* spp.), all four living species of which are considered threatened, and the odd-toed ungulates or perissodactyls (the horses, tapirs, and rhinos). Amongst birds, the albatrosses, bustards, cranes, parrots, pheasants, and cracids (the family Cracidae, comprising the curassows, guans, and chachalacas of the American tropics) all have a disproportionately high number of threatened species, including two of the largest flying birds —the magnificent wandering albatross *Diomedea exulans* and the great bustard *Otis tarda*— as well as some of the most beautiful of all wild creatures, such as the hyacinth macaw *Anodorhynchus hyacinthinus* and western tragopan *Tragopan melanocephalus*.

Reptiles

Fewer than one in six of the approximately 8 000 living species of reptile have been evaluated for their conservation status. Of these, roughly one in four —some 300 species in total— have been classified as threatened. It is not clear whether this figure can justifiably be extrapolated to the group as a whole. If so, this would indicate that overall reptiles are at least as threatened as mammals, and possibly more so. However, most of the groups that have been the focus of attention —crocodilians, chelonians (turtles, tortoises, and terrapins), and a few groups of lizards and snakes, such as the West Indian iguanas and the Eurasian vipers— are known to be experiencing specific threats that may not apply with the same intensity to reptiles as a whole. Crocodilians and chelonians in particular, the latter especially in East and Southeast Asia, have been subject to very high levels of exploitation which have contributed to their threatened status. Species such as the Bornean river turtle *Orlitia borneensis* and the Burmese roofed turtle *Kachuga trivittata*, considered safe a few short years ago, are now both classified as Endangered as a direct result of uncontrolled international trade.

Amphibians

Of the nearly 5 000 living species of amphibian (frogs, toads, salamanders, and newts), considerably fewer than 1 000 were assessed for the *2000 IUCN Red List*, and just under 150 species —around 20% of those assessed— are listed as threatened. The small number of species listed almost certainly gives a false impression of the plight faced by this group around the world. Virtually in every place where they have been studied, amphibian populations are declining, and in several parts of the world, in Central America and Australia for example, species have rapidly and inexplicably disappeared. Among those that may already be extinct are the spectacular golden toad *Bufo periglenes* of Costa Rica and the bizarre Conondale gastric brooding frog *Rheobatrachus silus*, an Australian species that rears (or perhaps reared) its young inside its stomach.

An international network of scientists comprising the Declining Amphibian Populations Task Force, established under the auspices of the SSC, has been investigating these disappearances and declines and identifying the range of factors suspected to be behind them, including chemical contaminants, introduced species, and disease. Threatened species assessments for individual species continue to be conducted based on this ongoing work, so this list is growing. That the current listings in the Red List severely underrepresent the degree of threat facing amphibians is also supported by the results of a comprehensive assessment in the United States by The Nature Conservancy showing that a far higher percentage of amphibians (36%) was threatened than either birds (14%) or mammals (16%).

Fishes

The conservation status of the vast majority of the world's 25 000 or so species of fish remains unknown. Those assessments that have been carried out, however, give grave cause for concern. Even with the limited effort expended to date, over 750 threatened fish species have been identified, the great majority in fresh waters. Many of these are important fisheries species, such as the sturgeons and paddlefishes (order Acipenseriformes), of which no fewer than 25 out of a total of 27 species are considered threatened. Other noteworthy threatened fishes include a number of species confined to single-cave systems, for example the Critically Endangered Namibian cave catfish *Clarias cavernicola*, or to desert springs, amongst which can be counted no fewer than fifteen threatened species of desert pupfish *Cyprinodon* spp. in northern Mexico and the southwestern U.S.A.

While most efforts to date have concentrated on fishes that frequent fresh waters, increasing attempts are being made to assess the status of marine species. Elasmobranchs —sharks, skates, and rays— have been a particular focus of attention. Of the approximately 1 000 species, 39 are currently listed as Threatened, with another 19 listed as Data Deficient; no fewer than 95 —or almost 10% of the entire group— are of conservation concern, and are expected to be listed once the assessment process has been concluded. Also heavily represented are a number of restricted-range coral-reef fishes, sea horses, groupers, and wrasses.

Invertebrates

Assessment of the status of invertebrates has come a long way since the pioneering efforts of the *1983 IUCN Invertebrate Red Data Book*. In a few cases, entire species groups have been covered, most notably the swallowtail butterfly family or Papilionidae, and efforts at assessing the status of the world's terrestrial and freshwater molluscs have made great strides. There remain, however, enormous gaps in knowledge. By and large, the only regions with even a remote understanding of the status of their invertebrate fauna are North America, Europe, Japan, Australia, New Zealand, and South Africa. Even here, only the more conspicuous and better-known groups, such as butterflies, dragonflies, larger beetles, snails, and large crustaceans, have been assessed in any detail. Clearly, the 2 000-odd threatened invertebrates listed in the *2000 IUCN Red List* represent only the tip of a very large iceberg. Among them are such bizarre and fascinating creatures as the wetas *Deinacrida* spp. of New Zealand, among the world's largest insects, of which three species are considered Vulnerable, and the magnificent, Endangered, Queen Alexandra's birdwing butterfly *Ornithoptera alexandrae* of Papua New Guinea. Of the few marine invertebrates listed, three vulnerable species of giant clam *Tridacna* stand out: these species are heavily exploited for use as food, and are evidence that unsustainable harvest of marine resources does not apply only to fishes and other vertebrates.

Plants

Some 5 600 plant species are listed as threatened in the *2000 IUCN Red List* (see Table 10). The great majority of these are trees —the result of a major global effort to identify threatened trees carried out by the World Conservation Monitoring Centre in collaboration with the SSC during 1995-1998, and published as *The World List of Threatened Trees*. Despite the large scale of this effort, only a small proportion of the world's trees were assessed, thus making it difficult to draw general conclusions from the results. Although over half the trees assessed in the course of the project were identified as threatened, assessment concentrated on those areas, such as the Southeast Asian rainforests, where tree populations were known to be under considerable pressure; hence the figures may not be representative of plants as a whole. The status of conifers may give a clearer indication of the overall picture, this being the only major plant group to have been comprehensively assessed using the new IUCN categories. Of a total of just under 900 species, some 140, or 16%, are currently considered threatened, a proportion that is roughly in line with the global figures for birds and mammals.

One innovation in the *2000 IUCN Red List* is the inclusion of around 100 species of bryophytes (mosses and liverworts), the first time that any members of this important but often neglected group of plants have been included in a global Red List. A number of carnivorous plants are also included, amongst them the extraordinary, Endangered Rajah pitcher plant *Nepenthes rajah* of Mount Kinabalu on Borneo, whose football-sized pitchers are reputedly capable of trapping animals the size of a squirrel. The *1997 IUCN Red List of Threatened Plants* includes representatives of no fewer than 372 plant families, including just under 1 800 species of orchid and nearly 600 species of cactus. A number of plants in these two families, including the Mexican cacti of the genus *Ariocarpus* and the fabled Rothschild's lady's slipper orchid *Paphiopedilum rothschildianum* (another Bornean species) are threatened by overcollection, commanding high prices amongst unscrupulous plant collectors.

The distribution of threatened species

Because information on threatened species in most groups of animals and plants is so incomplete, it is difficult to construct an accu-

to page *182* →

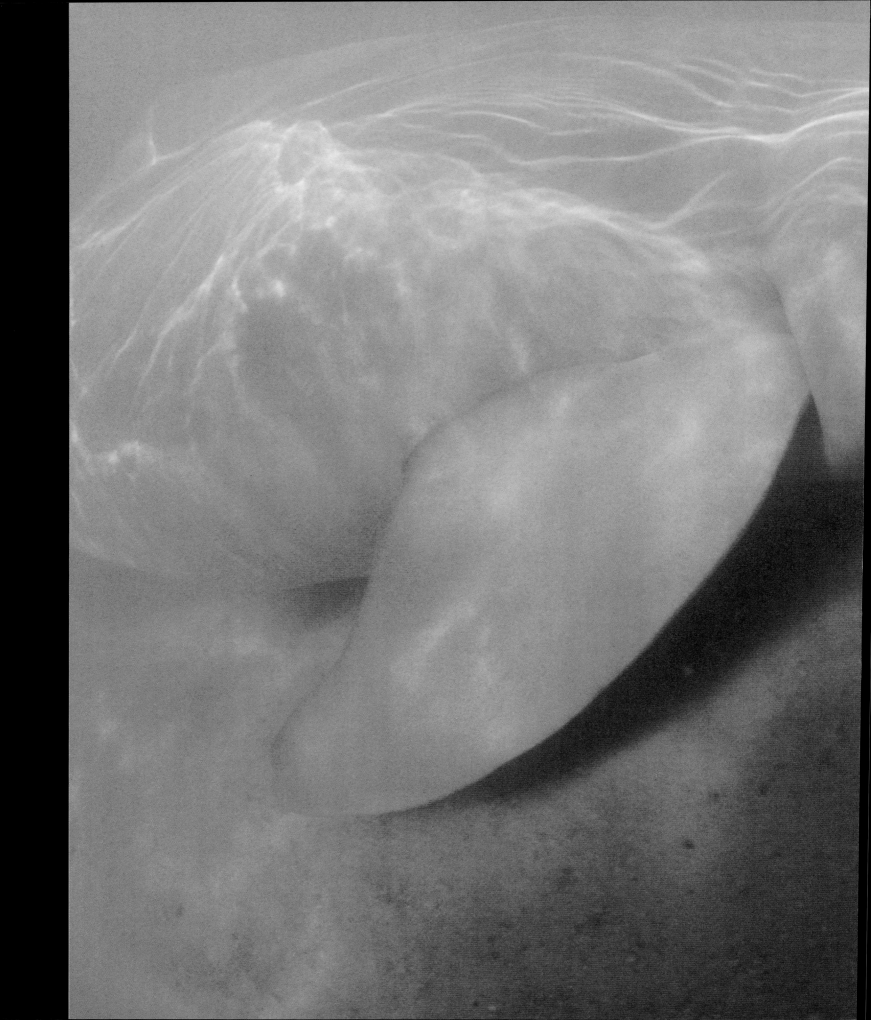

rate picture of their global distribution. Some indication, however, of those areas that are likely to have the greatest numbers of threatened species can be gleaned from analysis of the information on birds and mammals, where global coverage is reasonably complete. Of those countries that harbor the largest numbers of threatened birds and mammals, one —Indonesia— tops both lists. Other countries that feature prominently in both lists are: Brazil (second for birds and third for mammals), China (fourth in both lists), and India (sixth for birds and second for mammals). This is not particularly surprising: all four countries have extremely high levels of biological diversity, and all are experiencing tremendous environmental change.

However, these figures do not show the whole story. In terms of the proportion of bird and mammal species under threat, these countries are, surprisingly, not exceptional. In each case, threatened birds comprise about 5-8% of the breeding total, and threatened mammals 15-25% of the total number of species present. These figures are in line with many other countries, large and small. Those parts of the world with the highest proportions of their terrestrial faunas under threat are island nations such as New Zealand, the Philippines, and Madagascar.

Threatened terrestrial species

Given that on land most species occur in forests, particularly tropical rainforests, and considering the rate of destruction of these forests, it is to be expected that most threatened species would also occur there. Analysis of the habitat requirements of virtually all the threatened birds and around half the threatened mammals indicates that this is indeed the case. More than 900 species of threatened birds, over 80% of the total, use tropical rainforests, and over two thirds of these appear to be completely dependent on this ever-shrinking habitat type. Noteworthy are the Southeast Asian hornbills such as the Critically Endangered Sulu hornbill *Anthracoceros montani* of the Philippines and the cracids (curassows, guans and chachalacas) of the Americas, of which fifteen species are threatened, including the spectacular and Endangered horned guan *Oreophasis derbianus* of Mexico and Guatemala, and the Critically Endangered Peruvian white-winged guan *Penelope albipennis*.

While the plight of rainforests commands a great deal of public attention, species in other habitats may be under equal, or sometimes even greater, pressure. Grassland species have suffered from agricultural intensification in many parts of the world, particularly

North America and Europe, but increasingly also in the tropics. The species affected occur in many different groups, from orchids, including the North American Eastern prairie fringed orchid *Platanthera leucophaea* to vipers, such as the European Orsini's viper *Vipera ursinii*. On the Indian subcontinent, several of the most highly-threatened birds are grassland species such as the spectacular floricans, members of the bustard family, known for their magnificent aerial courtship displays. The extraordinarily diverse flora of areas with a Mediterranean climate has also suffered from clearance for agriculture, industry, and tourist development. Many of the plants from these regions have naturally small ranges and are thus inherently at risk. The richest of these areas —the Cape Region of South Africa— has seen several of its most spectacular species, such as the beautiful heather *Erica verticillata*, already extirpated from the wild, with many others reduced to a mere handful of individuals.

Whereas true deserts and mountain regions have suffered less from habitat conversion than other terrestrial ecosystems, they include some of the world's most threatened species, often victims of uncontrolled hunting. In the Sahara —the world's largest desert—, such uncontrolled hunting has driven at least one species, the beautiful scimitar-horned oryx *Oryx dammah*, to extinction in the wild and has critically imperiled several others. In the high mountains of Asia, hunting of the Tibetan antelope or chiru *Pantholops hodgsoni*, famous for its luxurious wool, continues unabated and the species is now considered Critically Endangered. Similarly, years of persecution along with a continuing reduction in its prey base have resulted in an endangered classification for the snow leopard *Panthera uncia*, perhaps the most beautiful of all the spotted cats.

Threatened freshwater and amphibious species

Although documentation of the threatened status of aquatic species is far less complete than for their terrestrial counterparts, as is the case with recent extinctions, available information suggests that freshwater species as a group may be considerably more threatened. Amongst the mammals and birds, for example, there are some striking examples: four of the five strictly freshwater cetacean species, the river dolphins of Asia and South America, are threatened, including the Critically Endangered baiji *Lipotes vexillifer* of China, now down to a few tens of individuals. Amongst the grebes, an ancient group of specialist diving birds inhabiting freshwater lakes and marshes, four of the 19 surviving species in the family are threatened, including one that is likely to already be extinct. Nine of the 15 species of cranes, among the most majestic of the world's birds, and closely associated with wetlands, are threatened with extinction.

The results of comprehensive assessments of threatened species in the U.S.A. conducted by The Nature Conservancy and featured in the Red List also point to freshwater species as being subject to the most serious threat; the highest proportion of threatened species were in these groups, for example, nearly 40% of freshwater fishes and a staggering 70% of freshwater mussels. The *2000 IUCN Red List* includes no fewer than 436 threatened freshwater molluscs, along with a further 90 that are classified as Extinct or Extinct in the Wild —ample evidence of an ongoing extinction crisis. Among the threatened freshwater invertebrates are such commercially valuable animals as the European freshwater pearl mussel *Margaritifera margaritifera*, classified as Endangered, and the Vulnerable noble crayfish *Astacus astacus*, another European species.

The alarming trends for freshwater fauna are linked to a wide range of human-induced impacts: sedimentation and organic pollution from land-based activities; toxic contaminants from industrial and municipal sources; stream fragmentation and flow regulation by dams, channelization, and dredging projects; bycatch in inland fisheries; and the introduction of exotic species.

Threatened marine species

One of the major contributions that the Red List has made over the past five years has been in highlighting extinction risk in marine species. For centuries past through to this day, prevailing scientific wisdom —and public perception— has been that marine species in general are resilient to extinction, owing to wide distributions through vast expanses of ocean and biological traits that include a high number of young to ensure their survival, even under heavy levels of fishing pressure. These perceptions —and our generally poor knowledge of the marine environment— meant that only a relatively small number of marine species have been thought to be threatened. The Red List (as well as other threatened species lists) has focused almost exclusively on marine mammals such as monk seals, sea otters, cetaceans, and sirenians, and marine turtles; a few high-profile fishes such as the totoaba *Totoaba macdonaldi* of the Gulf of California and the coelacanth *Latimeria chalumnae* of the Comoros and now Indonesia; and a few others, such as the giant clams of the family Tridacnidae.

However, a growing body of data from research in and on the oceans, combined with advances in scientific understanding of the extinction process and in the Red List system, have enabled scientists to test many assumptions about marine species, with surprising —and worrisome— results. As the Red List assessment process extends to marine species other than mammals, birds, and marine turtles, in particular marine fishes, many more species are being identified as threatened. In a major development, the *1996 IUCN Red List* included as threatened for the first time over 100 marine fishes assessed according to the new Red List system. A number of these were restricted-range coral-reef fishes, but others were commercially valuable fishes subject to fisheries such as the southern bluefin tuna *Thunnus maccoyii* and Atlantic cod *Gadus morhua*, which were listed on the basis of the IUCN decline criterion. Although the accuracy of some of these assessments, in particular for the latter, highly fertile species, has been questioned (so that these listings have been suspended, at least for the time being), overall this development proved a major breakthrough in thinking about extinction risk in marine species. Further, it prompted a range of studies and initiatives on the subject, which continue to yield surprising results.

While the listings of many of the fish species point to the impact of unsustainable fisheries, the Red List process is providing evidence of threats to marine species from other factors. Extensive changes were made in the threatened classifications of albatrosses and penguins in the *2000 IUCN Red List*, for example. A major threat for albatrosses was identified as incidental take in longline fisheries, while the threats to penguins include pollution, bycatch in fisheries (especially gill nets) and, for the more tropical species, declines associated with the repeated El Niño events, which have affected their prey base.

From listing to action

The *2000 IUCN Red List* and its antecedent and companion volumes, such as BirdLife International's *Threatened Birds of the World* (2000) and the *1997 IUCN Red List of Theatened Plants*, represent remarkable achievements. By themselves, however, they will do nothing to stem the tide of extinctions that threatens to engulf us. Rather they serve, as was their original intention, as a call to arms. In the following chapter, we will look at some of the actions to save threatened species that have already been prompted by the Red List and Red Data Book programs, and will examine what more needs to be done if the catalog of extinct species is not to continue to grow remorselessly.

ON HUMANS AND AFRICAN ELEPHANTS

For many people working in the field of species conservation today, the future often appears bleak, but this judgment must not be accepted at face value. There have been notable successes in our lifetimes, and there will be more in the future. However, before we can move forward with confidence, we should always go back in time and retrace our steps in order to fully understand past challenges and how they have been met, and take stock of the approaches that may represent the best hope for the future. The past can not always prescribe future actions, but it *can* bring lessons to bear and act as a compass, helping to point us in the right direction.

The African elephant provides a striking example of the current dilemmas faced in the conservation of species of singular concern, especially the charismatic megafauna cherished by so many in our modern world. Their story has not been only one of doom and gloom, but a patchwork of successes and failures across their extensive range. In several portions of their range, over the past 100 years, African elephants have experienced periodic episodes of near disastrous declines interspersed with marked improvements in their status. In some stages in the past, their circumstances were dramatically improving in certain parts of their range, while at the same time catastrophic declines were taking place in other areas. This legacy of good and bad times has left the species

with an uneven status across its vast range, which still spans the eastern, western, central, and southern corners of an immense and diverse continent. While, for the most part, these variations in circumstances require very different conservation strategies on the ground, there is also a strong synergy created through the search for common solutions to common challenges wherever and whenever possible. This synergy, born of shared vision, must grow if elephants are to survive the new millennium. Few species have as many issues with which to cope, their highly-valued tusks and meat placing them under extreme threat, and their large size and long lives putting them in direct competition with their human cohabitants.

The story of the African elephant provides additional invaluable insights into both the past and the future. Though little is known of their numbers and distribution prior to the last century, by the late 1800s, elephant numbers in Africa had been dramatically reduced by the world demand for ivory, a demand that at times seemed insatiable. However, as they entered the twentieth century, elephants in many parts of their range found themselves on very fertile grounds, as human populations had been reduced dramatically following the decimation of domestic stock across the continent, a result of the accidental introduction from Asia of a viral disease to which they had no immu-

nity. Elephants, in lands unoccupied by people, began a steady recovery. Yet before long, human populations exhibiting unprecedented rates of growth began to catch up. In the face of greater human densities and rapid land conversion, elephants were increasingly forced to seek refuge in the newly-created network of parks and protected areas that had sprung to life in many areas of the continent during the colonial era. Their growing densities, increasingly confined to these safe havens, wreaked havoc on the plant and animal diversity of many of these areas.

However, before these impacts were even noticed, let alone understood, elephants took their next big blow: a resurgence in the demand for ivory, which accompanied the expansion of the world economy throughout the 1950s, 1960s, and 1970s. This scourge continued for many years and resulted in dramatic declines of elephants throughout much of their range. In the same distant countries where greed for ivory had sounded the death knell for hundreds of thousands of elephants, there came a growing awareness of the impact of this trade on elephant populations. In the end, it was primarily the will of the people of these same countries that brought a sudden end to the trade, both legal and illegal, in "white gold" in 1990. Since that time, in many parts of their range, elephants began a slow recovery only to be met by another unrelenting reality. During the decades of elephant population declines, human populations had continued to grow. Recovering elephant populations found their favorite haunts now firmly settled and their former habitats altered forever.

Although the illegal killing of elephants for ivory and meat continues in some places, it is much reduced. There remain areas of the continent where a great deal more must be done to secure elephant populations from these threats. But today, when we speak of our desire for the recovery of elephant populations, we can not realistically talk of their coming back to 1900s levels because there is no room left for them to reinhabit. Approximately 80% of known elephant range is outside the parks and protected areas of Africa. For the most part, this same land is needed by Africa's burgeoning human populations. As in most situations of competition, there will likely be winners and losers. Our challenge is to try to design strategies and approaches that result in "win-win" situations whenever and wherever possible. But this outcome is easier to describe in theory than to achieve in reality.

Africa is the world's poorest continent. It is a land rapidly filling with people, and their numbers are steadily increasing. Africa epitomizes the harshest challenges of our modern world: poverty, starvation, war, socioeconomic injustice, corruption and, most devastatingly, the killer disease AIDS, which is silently and relentlessly removing much of Africa's unrealized hope for the future. It is in this setting that we must strive to conserve African elephants for their own sake and for the sake of all those people who believe that it

would be a much poorer world without them. But who will bear the cost of their survival? Is it the people who most cherish them, many living far removed from the realities of Africa, or those who must remain with the elephants, sharing their increasingly crowded and impoverished land on a day-to-day basis?

It is the people of Africa who will ultimately determine the fate of Africa's elephants. In order for them to continue to live side by side with the largest of land mammals and suffer the costs of their depredations on lives and property, there must be some benefit. It can not be the elephants, alone, that emerge the winners. Likewise, those of us devoted to their conservation can only hope that as the battle wages on we will not be left with only people in a land devoid of elephants. We all agree that this would be an impoverished world. So we must aim to strike a balance. The question is —how?

This challenge, which now confronts elephants throughout their range, is perhaps their most serious yet. Herein is a common problem that requires a common solution or, at the very least, a synergy created by people working together for the same cause. How can we shift the balance from elephants extracting a toll to elephants bringing a benefit? There are many government authorities, nongovernmental organizations, and committed individuals working hard to find or create this balance. Perhaps the most notable victories for elephants, to date, have come through the increasing development of common understanding born from a process of active dialogue. It is one of the most unique and valued aspects of human

nature and reasoning that once we begin to understand the perspectives and experiences of others, we can begin to work towards solutions together.

For better or worse, the future of the African elephant now depends on this unique characteristic of humans and the subsequent efforts of people to work towards sustainable, lasting solutions. If these are not found, the world will become a much poorer place, but if they *are* found, it will be a shared victory for both these highly-evolved and highly-valued species: *Homo sapiens* and *Loxodonta africana*.

HOLLY T. DUBLIN
IUCN/SSC African Elephant Specialist Group

AVERTING EXTINCTIONS:
PROSPECTS FOR THE FUTURE

In early 1951, an expedition was mounted by a band of ornithologists to a series of tiny islets in the mouth of Castle Harbour at the eastern end of Bermuda. Their aim was to find the nesting sites of an almost mythical bird, the cahow or Bermuda petrel *Pterodroma cahow*, a species which, until a handful of specimens had turned up in the early decades of the twentieth century, had been considered extinct for nearly 300 years. The expedition succeeded, but it was quickly realized that the species had been rediscovered only just in time. A mere eighteen breeding pairs remained, hardly any of which were successfully raising young. Having been unexpectedly rediscovered, the cahow seemed about to disappear a second time —for good.

The immediate threat was soon established. Destruction of the cahow's original nesting habitat in soft soils on the main islands of Bermuda, along with introduced predators, had forced the birds to nest on cliffsides favored by the white-tailed tropic bird *Phaëthon lepturus*, and the larger, more aggressive tropic birds were displacing cahows from their nest sites, almost always killing the latter's chick (female cahows only lay one egg per year) in the process. An ingenious solution was quickly found in the form of artificial nest entrances placed over the cahow's nest sites, which were large enough for the cahows to fit through but too small for the tropic birds. Mortality from tropic birds promptly ceased, and the cahow population began to increase.

Almost as soon as this problem was solved, another manifested itself in the form of a sharp decline in the number of eggs that were hatching. Analysis of unhatched eggs revealed high levels of DDT and other organochloride pesticides, which had been implicated in the drastic declines of bird-of-prey populations in Europe and North America. There were no local solutions to this problem, and for a while the future of the cahow again looked bleak. Fortunately, the imposition of major controls on the use of these pesticides in response to political pressure in Europe and North America, much of it prompted by Rachel Carson's 1962 classic *Silent Spring*, meant that by the late 1960s cahow breeding success had risen again.

Slowly the cahow population increased, though not without further setbacks, each of which had to be dealt with in turn. Several low-lying nests were regularly lost in high tides, and so, a seawall was constructed to protect these sites. The number of breeding-age pairs of cahows began to exceed the available nest sites, and so, artificial nest burrows were put in place. In 1987, bright security lights installed on a nearby U.S. Naval Air Station caused severe disruption to the birds, threatening to stop breeding entirely. Patient negotiations resulted in the lights' being switched off in late 1990, at which point breeding success immediately picked up again.

Thanks to these and other efforts, the cahow population has grown to the point at which nearly 60 pairs now nest each year. Many organizations and individuals have played their part in bringing the bird back from the brink of extinction, not least the Bermudan authorities who had declared the cahow's nesting islets and surrounding waters a national park soon after the rediscovery of the species. However, the fact that the cahow is still with us is very largely due to the untiring efforts of one man, Bermudan naturalist David Wingate, who took on full-time responsibility for the species as long ago as 1958 and carefully nurtured the breeding population since that time.

The lessons of the cahow story are manifold. It shows that a handful of individuals, or even a single individual, can genuinely make a difference to the survival of an entire species. It also demonstrates that the price of successful conservation action is eternal vigilance: threats to species may be far from obvious at first, and it is hard to predict when a new one will arise, or what it will be. And it shows that while some problems may have surprisingly simple, local solutions, certain major threats to biodiversity can not be tackled at a local level, but rather call for concerted global efforts to be solved.

The cahow is far from being the only animal whose status has been turned around. In South Africa, two antelope species, the bontebok *Damaliscus pygargus* and black wildebeest *Connochaetes gnu*, were —according to Dr. Rod East, Cochairman of the IUCN/SSC Antelope Specialist Group— within a "whisker of extinction" last century, but both have now recovered to such an extent that they are listed as Lower Risk in the *2000 IUCN Red List*. The recovery of the southern population of the white rhinoceros *Ceratotherium simum* has been even more dramatic. At the end of the nineteenth century, a mere 10 to 20 animals of this magnificent species survived in the whole of the southern African region, in and around Umfolozi Game Reserve in KwaZulu-Natal, South Africa. Under the management of the Natal Parks Board, the population increased to levels which, at the beginning of the 1960s, made it possible to begin translocation of surplus animals to reestablish populations in the rhino's former range in southern Africa and to create, as insurance, new populations in Africa and elsewhere. So successful has "Operation Rhino" been that there are now over 10 000 white rhinos in nearly 250 separate populations, making it by far the most abundant of the five living rhinoc-

eros species. Like the bontebok and the black wildebeest, it too is currently classified by IUCN as Lower Risk. In stark contrast is the last remaining northern population of white rhino, which clings on precariously in Garamba National Park in the Democratic Republic of Congo, where the population of fewer than 35 individuals is under constant threat from poachers.

Endangered plants have also benefited from such dedicated conservation action. The Round Island bottle palm *Hyophorbe lagenicaulis*, a striking ornamental species endemic to Round Island in the Mascarenes had, by the early 1980s, been reduced to eight mature individuals and a single clump of six young trees, under severe threat from introduced rabbits. After much effort, rabbits were finally eliminated from the island in 1986, enabling recovery of the devastated natural vegetation to begin. Regeneration of the palm was actively assisted by botanists and SSC members Wendy Strahm and Ehsan Dulloo, both of whom dispersed seeds widely around the island. Today, there are hundreds of bottle palms growing in the wild, and the population is well on the way to recovery.

It would be gratifying to be able to say that each of the threatened species in the IUCN Red List has been the focus of such concentrated and successful efforts. Sadly, this is far from the case. The number of species that are the subject of active, intensive conservation plans, though impressive, is a tiny fraction of those deserving such attention. Closing this gap requires radical —and urgent— effort, buttressed by general approaches that will benefit many species at one time.

The need for multi-pronged conservation strategies

Just as most species suffer from a range of threats, often acting synergistically, effective conservation strategies must incorporate a wide range of measures to be taken at different levels: local, national, regional, and global. These strategies also necessitate a capacity to adjust measures in response to changing circumstances. Finally, they require a constant input and analysis of data on the species concerned, their habitats, and the threats themselves.

The weapons in the conservation "arsenal" are many and varied, and include: the establishment of protected areas and protected-area networks; broader habitat conservation in the form of management and restoration; prohibitions or restrictions on exploitation and the use of pesticides, herbicides, and other chemical pollutants; active eradication programs for invasive species; captive breeding and reintroduction; and training and technical capacity-building for species

to page *220* →

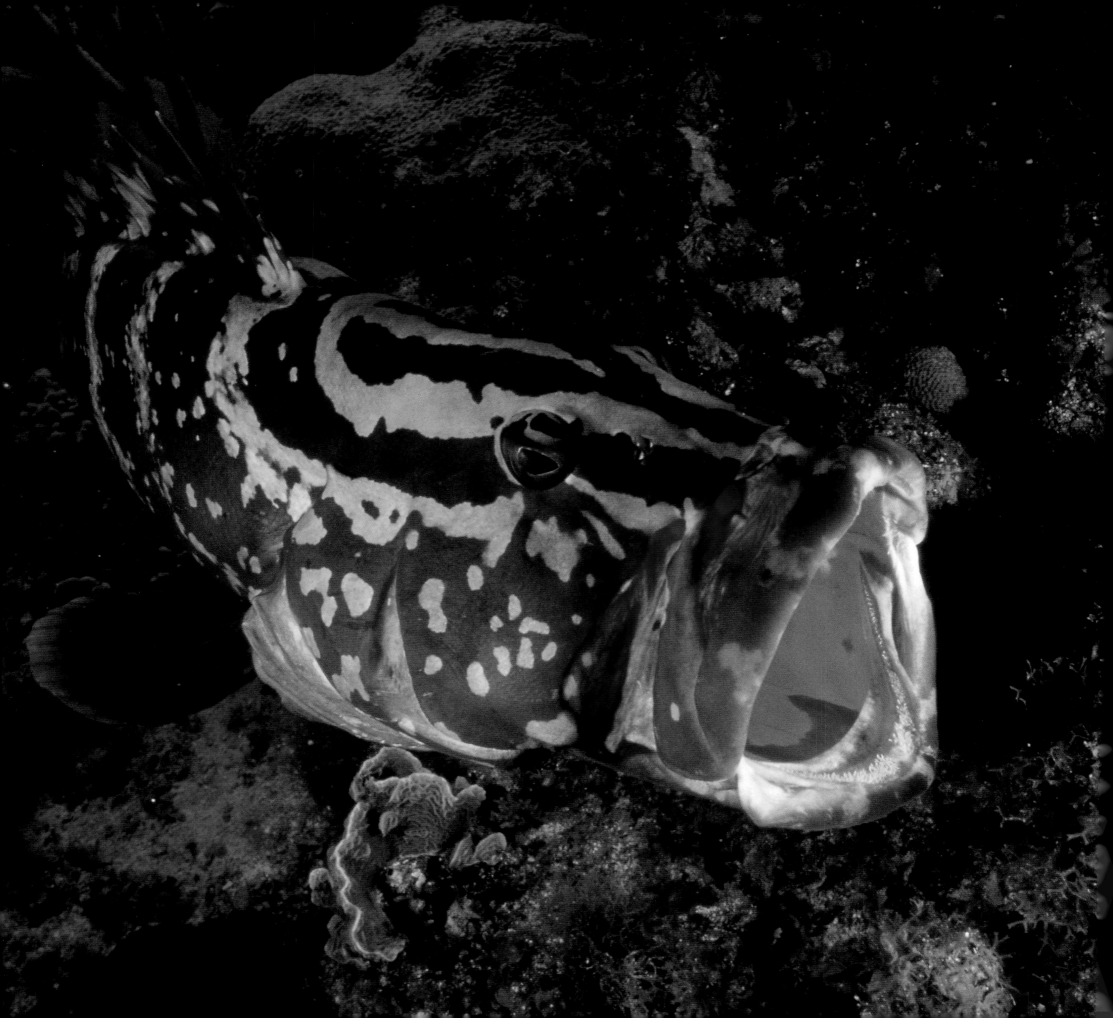

and biodiversity conservation and environmental management. These measures require that adequate legal and regulatory frameworks be put in place; that management agencies be designated and provided with resources to meet their mandates; that research be undertaken to fill key gaps in knowledge about species and habitats and their responses to threats and management actions; and that local communities and other stakeholders be engaged to foster and, increasingly, participate in these efforts. In addition to human, financial, and technical resources, these strategies must rely on broad political support to be successful; developing that support is one of the most important activities to pursue in achieving biodiversity conservation.

The conservation community —including IUCN and its Species Survival Commission (SSC)— is constantly designing new tools to employ in conserving species and biodiversity. An important outcome in recent years is the application of expertise and methods from the nonbiological sciences such as sociology, anthropology, and economics, in understanding and addressing more effectively the dynamic between people and biodiversity. The development of stakeholder processes that involve all the interested parties in decision-making, as well as collaborative management strategies which vest authority and responsibilities in different stakeholder groups, have been particularly promising in this regard, as have efforts to establish alternative sources for essential materials from threatened species and markets for ecologically-based materials such as nontimber tropical forest products and goods made from them. But there is always a need for more tools, particularly as new threats emerge or existing threats intensify and become more complex. This is one of the challenges of effective conservation: a constant need to learn, adapt, and involve others in the process.

Protected areas

Since most Red List species are currently threatened either by habitat loss or overexploitation or, often, some combination of the two, the most straightforward means of conserving them, at least in the medium term, is by setting aside areas where both they and their habitats are protected. National parks, game reserves, and other forms of protected areas are nothing new —areas of forest have been protected in India, for example, since the fourth century BC, and in Europe recreational hunting reserves have a history stretching back centuries. More recently, the setting aside of natural or seminatural areas for the common good began in earnest in the late nineteenth

century and has progressed to the point that most countries in the world have a protected area network of some kind.

Increasingly, protected area systems are incorporating freshwater and marine habitats as well. In Japan, the Dragonfly Kingdom at Nakamura, which supports more than 80 dragonfly species, is one of over twenty reserves set up in the country to conserve these spectacular insects and other aquatic species. Elsewhere in East Asia, a major project is being developed by the International Crane Foundation, the governments of China, Iran, Kazakhstan, and Russia, and others to protect a series of critical wetlands along the eastern and western flyways of the Siberian crane *Grus leucogeranus* that are also key habitats for many other migratory water birds. Major initiatives are under way to expand the coverage of marine protected areas (MPAs) in order to conserve marine species and habitats and, in some instances, enhance populations of economically important marine resources. No-take zones, or marine reserves, which serve as refuges from direct and indirect effects of fishing, can be instrumental in conserving species and habitats and can help to replenish stocks outside these areas. In both developing and developed countries, they are widely gaining acceptance as both a conservation and fisheries management tool and are being employed as an effective, low-cost, "low-tech" means to achieve these often-conflicting objectives. For some threatened marine species for which fisheries bycatch is the major threat, marine reserves may offer their only chance for survival.

Because of the way that wild species are distributed, even a small percentage of a region's area set aside may protect a surprisingly high proportion of the species present there. It has been estimated, for example, that well over 70% of sub-Saharan Africa's bird species are living in protected areas, despite the fact that these areas cover less than 5% of the region, while in southern Africa over 90% of mammal, bird, reptile, and amphibian species are represented by breeding populations in protected areas. Similarly, one protected area may provide refuge for a number of threatened species. In Rio de Janeiro, Brazil, the Desengano State Park holds populations of no fewer than 18 threatened bird species, while Ranomafana National Park in Madagascar is home to at least ten threatened mammal species, including two Critically Endangered lemurs, the golden bamboo lemur *Hapalemur aureus* and the broad-nosed gentle lemur *Hapalemur simus*. There are many other such areas around the world: Taï National Park in Côte d'Ivoire, Manú National Park in Peru, Royal Chitwan National Park in Nepal, El Triunfo Biosphere Reserve in Mexico. All these are of extraordinary importance for the conservation of biodiversity, and are the crown jewels of the natural world.

Despite these achievements, there are still many challenges in ensuring that protected areas save species. It does not suffice, for example, for a species to be present in a protected area; the area must be large enough to support a self-sustaining population of that species and, crucially, it must be protected in reality, not just theoretically. Many of the world's most important protected areas for threatened species are in developing countries, which have extremely limited resources to commit to protecting them. Moreover, they are often in areas with rapidly growing human populations, which places intense pressure on the land for conversion to agriculture and livestock grazing, and on natural resources such as timber and wildlife. Understandably, therefore, encroachment and poaching are rife in many protected areas.

Even where areas are adequately protected, they are often subject to a range of conflicting management interests. Many current protected areas have been established for reasons quite other than to safeguard species and their habitats. They may protect areas of outstanding scenic beauty or important archaeological and historical sites, or may be intended primarily to offer recreational opportunities for growing urban populations. Management for these ends may, potentially, conflict with the needs of wild species (as well as with each other).

Moreover, it is often impossible to protect the entire area over which one population and sometimes even one individual ranges. For many migratory or wide-ranging animals, particularly birds and many marine species, protection of particular critical habitats —seaturtle nesting beaches, overwintering sites of migratory water birds, or spawning sites of groupers and other commercially valuable marine fishes— may be pivotal to their survival.

Finally, it is almost impossible to keep protected areas entirely isolated from external influences. On land, fences may control human access and exclude some undesirable animals, but they can not prevent the impact of airborne pollution, for example. In aquatic ecosystems, this problem is even more severe. Rivers and marine areas rely for their integrity on a constant flow of water. Any barriers that might be erected will disrupt water flows, with potentially disastrous consequences. Even when an entire enclosed water body, such as a lake, is included in a protected area, often the catchment that feeds it lies outside the area and may suffer from a range of adverse influences.

Conserving habitats outside of protected areas

While protected areas are, in the short term at least, probably the single most important weapon in the fight to save threatened species, they can not ever be the whole solution. For the reasons mentioned above, as well as the simple fact of human demographics and the associated increasing demand on land, water, and other resources, it is clear that important wildlife habitats can not always be protected or that such protection can not in and of itself guarantee species' survival. Other strategies must be implemented to conserve, through management and, in some instances, restoration, habitats outside of protected areas.

Just as biodiversity —and threatened species— are not evenly distributed, people are not spread out evenly across the planet. Be it in the U.S.A. or Tanzania, people are disproportionately found around coasts and in areas with good soils and other favorable living conditions; this is often the case for species as well. One recent study has revealed that important sites for biodiversity conservation in Africa, in particular those harboring a large number of endemic or threatened species, coincide with areas of high human density, thus suggesting that conservation efforts in those areas need to specifically (rather than obliquely) address managing people as well as species. While this can be in the form of protected areas, the rapid growth of human populations, and in many instances those of domestic livestock, necessitates more innovative approaches to balancing wild species and habitats with human needs and aspirations.

One strategy that is increasingly being pursued around the world is the development of biodiversity corridors. Linking protected areas or managing different protected and unprotected habitats as a larger unit allows for connectivity between these areas, thus minimizing the negative effects of habitat fragmentation, which is a major cause of species loss. This corridor or network approach can be more effective than establishing a single fixed-boundary protected area, and it can specifically incorporate human settlements or areas affected by human activities as part of an overall biodiversity conservation plan.

Captive breeding and artificial propagation

Where efforts at maintaining wild populations or controlling serious threats to them are failing, captive breeding (of animals) or artificial propagation (of plants) may be the only option to prevent extinction. This has been the fate of a number of species, some well-known, some obscure. Amongst the birds, the Guam rail *Gallirallus owstoni*,

to page *230* →

Socorro dove *Zenaida graysoni*, and the impressive Brazilian Alagoas curassow *Mitu mitu* have all become extinct in the wild, while amongst mammals, the list of species whose survival depends on captive breeding includes the North American black-footed ferret *Mustela nigripes*, the scimitar-horned oryx *Oryx dammah*, and the Saudi gazelle *Gazella saudiya*. To this list can be added a number of molluscs, including nine *Partula* snails from French Polynesia, several fishes, and even a few insects such as the Hawaiian Oahu deceptor bush cricket *Leptogryllus deceptor*. Captive breeding is also being pursued for many other threatened species whose status in the wild is quickly deteriorating, with active involvement of the world zoo community and the IUCN/SSC Conservation Breeding Specialist Group.

There are many quite notable successes of captive breeding of threatened species, but it is in few cases a simple and straightforward endeavor. Maintaining populations of animals in captivity over long periods is generally expensive and difficult. Some animals are simply very hard to keep alive, let alone breed, in captivity. Even if success can be achieved, if the initial or founder population is very small, as is often the case with highly threatened species, problems of inbreeding are likely to arise. Moreover, as it is very risky keeping all the remaining individuals of a species in one place, an effective captive-breeding program ideally involves long-term cooperation amongst a number of different institutions, something that is not always easy to accomplish. Because of these and other inherent difficulties, deploying captive breeding in conserving a species that is not already represented in captivity may be fraught with controversy. The black-footed ferret was almost certainly saved from extinction through captive breeding, but attempts to establish a captive-breeding population of the Critically Endangered Sumatran rhinoceros *Dicerorhinus sumatrensis* have been much less successful, thereby lending credence to the argument that conservation resources would be better directed to trying to maintain the last few natural populations in the wild. Whether or not to move animals into captivity is a common dilemma in conservation: these are rarely easy decisions.

Whatever the desirability or otherwise of captive breeding in particular cases, it is extremely unlikely that, in the near future at least, it will be a viable option for more than a small proportion of the world's threatened animal species. The case with plants is somewhat different. It is, on the whole, much easier and cheaper to maintain plants in cultivation than animals in captivity —evidenced by the vast range of species that are grown by millions of gardeners around the world. Indeed, some species that are extinct in the wild are widely grown as ornamentals, including the tree *Franklinia alatahama*, extinct in its native Georgia, U.S.A. since the nineteenth century; the Turkish tulip *Tulipa sprengeri*; and the fabled Chilean blue crocus *Tecophilea cyanocrocus*. However, as with animals, there are many plants that prove difficult to establish or maintain in cultivation. It is often problematic keeping those that *do* grow and reproduce well in truly wild form over long periods. Many species easily hybridize with closely-related species —this is the origin of countless cultivated varieties— and over a few generations in cultivation some plants may become very different from their wild progenitors. A further practical problem arises from the fact that most threatened plant species occur in the tropics, while most botanical gardens and other institutions specializing in the cultivation of plants are in developed countries in temperate parts of the world. While smaller tropical species, such as many orchids, can be maintained under glass in such institutions, it is scarcely practicable for viable populations of the large number of threatened tropical trees to be maintained under these circumstances.

Captive breeding and artificial propagation can also play a valuable role in production for commercial markets. Be it parrots for the pet trade or plants for the horticultural or pharmaceutical trades, commercial production in captivity, or *ex situ*, can alleviate pressure on wild populations. It doesn't necessarily do this, however. Careful planning, and often pricing, are required, as well as monitoring of the trade and its effects on wild populations, in order to verify and ensure that commercial production *ex situ* meets conservation objectives.

Reintroduction

For almost anyone with an interest in the natural world, the survival of species only in captivity or cultivation is a compromise measure at best. The desirable aim should ultimately be to reestablish the species in its natural habitat. Throughout the world, captive-breeding operations are being undertaken to meet this specific objective —as an integral component of a broader strategy to reestablish or reinforce species' wild populations. Hence, Przewalski's horse *Equus ferus przewalski* is being reintroduced in Mongolia and China with the assistance of the IUCN/SSC Equid Specialist Group, and wild populations of sandhill cranes *Grus canadensis* and whooping cranes *Grus americana* in the U.S.A. are being reinforced through release of captive-bred animals. Also in the U.S.A., the Critically Endangered American burying beetle *Nicrophorus americanus*, named for its practice of burying carcasses as food for its larval young, is being captive-bred and reintroduced to Nantucket Island, from where it disappeared in 1926.

In the Caribbean, populations of the Critically Endangered Jamaican iguana *Cyclura collei* are increasing thanks to an innovative program guided by the IUCN/SSC Iguana Specialist Group, whereby young iguana hatchlings are taken from the wild and raised at the Hope Zoo in Kingston, then released (with radio transmitters for monitoring) back into the wild. The results thus far suggest that this strategy offers serious promise for the future not only of this species but also for at least one other Caribbean iguana, the Critically Endangered Anegada rock iguana *Cyclura pinguis*, for which a similar "headstarting" program has been initiated.

So successful has reintroduction been as a conservation tool, and so obvious the lessons when attempts at reintroduction have failed, that IUCN, through the SSC, has produced *Guidelines for Reintroductions*, which provide "best practice" advice. The IUCN/SSC Reintroductions Specialist Group also provides guidance to individual efforts, including a practitioners' directory that has contact addresses for organizations and individuals involved in reintroduction projects. Several general rules apply. First, it is essential to identify —and control— the factors that led to the species' decline or disappearance in the wild, lest reintroduced animals and plants suffer the same fate, which would render the effort futile. Second, it is essential to conduct prerelease screening —for disease and other risks to wild populations— and postrelease monitoring so as to identify and address the positive or negative effects of the reintroduction. Finally, reintroduction programs require serious and sustained commitment if they are to succeed.

Managing very small populations

Once populations have been reduced to very low levels, as was the case with the cahow, simple protection of a species and its habitat may no longer be enough. Any factor, natural or otherwise, that affects the survival of individuals or reproductive success takes on greatly magnified importance, and dealing with it successfully may make all the difference between survival and extinction. Not only must there be active monitoring of these populations in order to detect, and in many instances predict, impacts, but a wide range of management activities may have to be undertaken. These might include habitat manipulation through controlled burning or flooding, translocation of individuals to repopulate parts of the species' origi-

nal range or move them out of problem areas, and active control of predators or competitors. Aside from headstarting, the recovery of the Jamaican iguana relies on a program to remove introduced mongooses, cats, and rats from the key iguana areas; protection and management of iguana habitat; and, eventually, the reestablishment of an iguana population on an offshore island that is part of the species' former range. This kind of ongoing, active management requires long-term commitment of humanpower and resources and the ability to respond quickly and decisively to changing circumstances.

The case of the Critically Endangered Western Australian swamp turtle *Pseudemydura umbrina*, one of the world's rarest reptiles, also amply illustrates these challenges. Known only from a few swampy areas in Western Australia, this small freshwater turtle had declined from a population of over 200 individuals in the late 1960s to around 30 by the mid-1980s, with only one viable population located in the 65-hectare Ellen Brook Nature Reserve. As with many Australian species, a major threat to the turtle was mortality from introduced predators, particularly foxes. In 1990, an area of 29 hectares within the reserve was fenced off to exclude these. Breeding success of the turtles rapidly increased, but soon they faced another problem. In the absence of predators, the nesting population of Pacific black ducks *Anas superciliosa* in the fenced area also increased greatly, but because of the fence, the flightless young ducklings were unable to disperse into adjacent ponds and reservoirs, as they would normally do, and large numbers built up in the swampy areas frequented by the turtle, dirtying the water, competing for food, and even attacking some hatchling turtles. The problem was solved by including within the fenced area an adjacent portion of land with a larger pond in it. The ducklings now move here until they are large enough to fly and no longer disturb the turtles. The fence also enclosed a substantial western grey kangaroo *Macropus fuliginosus* population that suffered high mortality in the years after fencing. This attracted large numbers of native ravens, which also attacked young turtles, particularly as the latter moved to higher areas where they spent the summer buried underground. This has been remedied with the help of volunteer raven-scarers who patrol the area at critical times. As Dr. Gerald Kuchling, Chair of the IUCN/SSC Madagascar and Mascarene Reptile and Amphibian Specialist Group, who has masterminded the recovery of this species over twenty years, points out, given these problems, it would be far simpler if the species had not been allowed to become so reduced in number in the first place.

to page *238* →

Seed banks and cryogenic "zoos"

Seed banks, where plant seeds are stored under carefully monitored conditions of temperature and humidity, can be a highly efficient and effective method of conserving plants away from their natural habitat. However, as with collections of growing plants, seed banks require constant maintenance and supervision, as well as secure power supplies to maintain the appropriate conditions for storage. Every batch of seeds must be tested periodically, and if the percentage that germinates falls below a certain level, and seed can not be collected to replace the older batch, seedlings from the original collection will have to be grown to maturity, a time-consuming and potentially expensive undertaking. Moreover, it is estimated that as many as 50 000 plants, including a large number of tropical species, produce seeds that have no natural dormant period and do not survive in seed banks.

The use of so-called "cryogenic zoos," in which eggs, sperm, and embryos are deep-frozen, is still a highly speculative and experimental endeavor. Although this technology is increasingly used in animal husbandry for livestock species, it is clear that it will be many years into the future before it plays a significant role in the conservation of biodiversity. The same can be said of cloning, which is increasingly suggested as a "high-tech" fix for saving critically endangered species. It will be many years yet before an assessment can be made of its prospective role in preventing extinctions.

National approaches

Great progress has been made in the development of national conservation legislation around the world, but there is still a great deal of work to be done to enhance the role of legislation and associated measures in protecting threatened species, managing species to prevent their becoming threatened, and conserving and managing habitats in and outside of protected areas. In many countries, for example, there is no standard process for protecting species as they become threatened, or the threatened species laws in place do not apply to certain species groups such as fishes, invertebrates, or plants. In order to be effective, legal restrictions must be appropriately framed and adequately enforced, but often this is not the case. Furthermore, it is clear that threatened species legislation, unless it is very comprehensive, can not address all the threats —real and potential— that are putting species at risk of extinction, in particular the presence of alien invasive species, pollutants, and the many forms of habitat destruction and deterioration. As many species featured on the Red List are restricted in range and limited to single countries, their chances for survival could be greatly increased through improved legislation and other such measures at the national level. Actively conserving species at the local and national level is the best insurance against global extinction.

One very promising outcome along these lines has been the development of national biodiversity strategies as called for under the Convention on Biological Diversity (CBD). In engaging a wide group of scientists, nongovernmental organizations, and government agencies in the effort, these strategies and action plans are identifying specific issues to be addressed and measures to be taken to conserve biodiversity in individual countries. More often than not, these strategies include the elaboration of threatened species lists and assessment processes that are drawing specifically from the IUCN Red List and the action plans and other expertise of the SSC. In tandem with the development of these strategies in many countries is the elaboration of biodiversity legislation to protect threatened species, establish protected areas, control the movement of potentially invasive species, and facilitate many other conservation measures. Once finalized, these biodiversity strategies must rely for implementation on adequate funding and a broad base of support from both private and public sectors.

International approaches

While for many species the chances for survival lie primarily with national solutions, increasingly there is a need for international approaches, as frameworks for measures to be taken at national or local levels or as a platform for collaboration and harmonization of approaches. Such has been the lesson of the Convention on International Trade in Endangered Species of Wild Fauna and Flora (CITES), for example, that in order for trade controls to be effective, they generally need to apply globally, lest trade shift from one country to the next. It has also been the case with the Convention on Migratory Species (CMS), which by its very nature focuses on united, and uniform, efforts to conserve species that cross national boundaries, and the Ramsar Convention on Wetlands, which has provided political impetus and technical support for the protection and management of wetlands and other aquatic habitats around the world. Numerous other international treaties have also been successful in fostering species conservation. Possibly the best known is the International Convention for the Regulation of Whaling, under which a moratorium on commercial whaling, established in 1985, has allowed some severely depleted populations of great whales, including the Endangered blue whale *Balaenoptera musculus*, the world's largest animal, to stabilize, and others to increase. IUCN and the SSC have been key players in the development and implementation of all of these treaties; in the case of CITES, for example, SSC, through the IUCN Wildlife Trade Programme, serves as the major source of technical information on species.

The situation of the polar bear *Ursus maritimus* offers perhaps the best example of the evolving role of international measures in addressing conservation issues. According to Dr. Stansilav Belikov and Dr. Scott Schliebe, Cochairmen of the IUCN/SSC Polar Bear Specialist Group, the 1973 Agreement on the Conservation of Polar Bears has been very successful in fostering cooperation between range countries on research and management of polar bears, in particular in relation to hunting. However, in recent years, studies on climate change and the occurrence of high levels of persistent organic pollutants in the Arctic have raised serious questions about the future of polar bear habitat and its implications for polar bear populations. Unlike overhunting, which can be addressed through relatively simple measures at national and regional levels, neither of these recent developments can be solved by managers on the ground, nor necessarily through the existing treaty. These emerging issues, which strike at the core of environmental integrity, require much more concerted, global approaches.

Another problem requiring worldwide measures is that of invasive species. The IUCN/SSC Invasive Species Specialist Group has been an active participant in global efforts to contain the threat of invasive species. The Specialist Group took the lead in producing *IUCN Guidelines for the Prevention of Biodiversity Loss Caused by Alien Invasive Species*, designed to assist countries, conservation agencies, and concerned individuals in reducing threats posed by invasive species, and is producing a global database and "early warning system" for alien invasive species, in collaboration with international scientific organizations, the CBD, and others.

Setting priorities

With so many threatened species and habitats to conserve and far fewer resources than needed being deployed to the task, governments, conservation organizations, and others are —understandably— exploring ways to achieve the most good in the most effective way.

One strategy for setting priorities has been to focus conservation efforts on those areas where the greatest number of species or threatened species occur. This "biodiversity hotspot" approach has been developed most extensively for terrestrial ecosystems and has resulted in the identification of 25 or so terrestrial "hostpots" around the world, most of them in the tropics and areas with a Mediterranean climate. These areas harbor the majority of the world's terrestrial species and, a recent analysis of the IUCN Red List has suggested, the majority of the world's threatened species. Although few would argue that virtually every natural habitat and wild species should not be conserved, a focus on these areas offers the hope of conserving a significant portion of the world's terrestrial biodiversity.

Although the IUCN Red List generally has no automatic effect —it is not a legal instrument—, it is an important tool in assessing the relative importance of species and threats for conservation action. At the global level, it is an essential reference for implementing international agreements such CITES and the CMS, both of which provide for measures to conserve the species listed in their annexes. By highlighting Critically Endangered species, the Red List can be used by any conservation agency or interested party desiring to focus on those species most severely at risk of extinction. In addition to other threatened species which can be prioritized according to the degree of threat, the Red List provides information on the threats involved, such as exploitation, disease, or alien invasive species, which can be used to prioritize conservation measures. Increasingly, aside from "hotspot" studies, it can be used for analyses that go beyond individual species, such as reviews of protected areas and the relative importance of different threats, including in different biomes. The evidence provided by the Red List, for example, of the extent and severity of extinction risk for freshwater species should provide particular impetus to efforts to conserve and restore freshwater ecosystems, while the role of fisheries bycatch in bringing several marine species close to extinction should also spur efforts to resolve that threat. At national and regional levels as well, the Red List is proving to be a valuable tool in priority-setting.

In addition to producing the Red List, the IUCN/SSC, as a whole and through its 130 taxonomic and disciplinary Specialist Groups, plays a significant role in priority-setting for conservation action. As an important complement to the Red List, the SSC produces status surveys and action plans for individual species groups. These analyze in depth the conservation issues facing different species groups

to page *250* →

and recommend specific measures as priorities for action, ranging from legal protection at the global level, such as through CITES, to working with one or more local communities to resolve conflicts with a very localized species. Moreover, these action plans serve as a virtual treasure trove of information to be used in the development and implementation of conservation projects and programs at various levels. The SSC and SSC Specialist Groups also assist priority-setting by providing technical advice and policy guidance through official policy statements and guidelines but also on a case-by-case basis to governments, nongovernmental organizations, local communities, businesses, and others on a range of issues from captive breeding and reintroductions to control of invasive species and appropriate restraints on trade in wild animals and plants. In the everyday work of monitoring trends in the status of plant and animal species and also of identifying and analyzing issues and solutions, the SSC serves as the single most important source of information and advice on species for the global conservation community.

Creating incentives for conservation

There are many practical challenges to conserving species today. However, the major obstacles to be overcome are not technical, but rather social, political, and economic. The maintenance of biodiversity, however crucial it is to our long-term well-being, is still accorded low political priority almost everywhere. This means that conservation efforts are typically underfunded and underresourced and, where there is conflict —over land use, for example, or the exploitation of species— conservation tends to lose out to other interests. Although a problem everywhere, it is particularly acute in the world's developing countries, especially the tropics, where most of Earth's biodiversity, and most of its threatened species, are found, because economic development and the alleviation of poverty are overriding national priorities considered (however erroneously) disjunct from or in direct conflict with species and biodiversity conservation.

This problem may be particularly keenly felt at the local level. Where habitats are set aside for conservation, or where stringent regulations are imposed on natural resource use, it is usually local people who must bear the cost. Much of this cost is in the loss of potential benefits if the land or other habitat were to be used in another way, in the case of a forest by being logged, for example, and then converted to agricultural or grazing land, or being developed for housing or commercial uses. Local people, particularly those living under a subsistence or near-subsistence regime, may not perceive any benefits from conserving a threatened species such as a frog or a butterfly. Redressing this imbalance, whereby often poor local people effectively bear the cost of maintaining a global benefit (or global good as economists would term it), is one of the greatest challenges facing conservation today.

Many different approaches are being explored to better accommodate the needs of both people and wildlife. In the U.S.A., where the major proportion of critical habitat for endangered species is land held in private hands, incentive schemes and education are key components being pursued by government and nongovernmental organizations to enhance the conservation of these species on private land. In many instances, private landowners may be able to manage their land, through reduced pesticide use or restoration of native vegetation, for example, in relatively simple, straightforward —and inexpensive— ways that dramatically improve its role in conserving species. Economic incentives, in the form of lower property taxes or reduced purchasing costs in return for forgoing development rights, are also being employed. In Europe, as well, farmers are being offered economic incentives to set aside some of their agricultural land as wildlife habitat. In southern Africa, private game farms are conserving species and providing landowners with options for generating income through tourism. In the case of predators and other species coming into conflict with humans, incentive schemes include compensation and other mechanisms to encourage farmers and others to tolerate these species —and their depredations— on their property.

One major innovation in improving the balance between people and wildlife has been the development of the concept of sustainable use. This argues that if people can derive some benefit from wild species or wild lands they are much more likely to try to conserve them for the future. The development of crocodilian ranching, much of it with the encouragement and guidance of the IUCN/SSC Crocodile Specialist Group, has undoubtedly improved the conservation status of several crocodilian species, while in Zimbabwe, the CAMPFIRE program has assisted local communities in setting up management regimes whereby they can derive income from exploitation of their wildlife resources. In Papua New Guinea, ranching of the magnificent, Endangered Queen Alexandra's birdwing butterfly *Ornithoptera alexandrae* has brought benefits to local peoples, while in South America, attempts are being made to revive the manufacture of vicuña *Vicugna vicugna* cloth, using wool sheared from free-living vicuña that occupy land in demand for grazing domesticated animals. All around the world, strategies are being developed to create incentives for native wildlife and habitats to be maintained rather than be replaced by agriculture or domestic livestock.

While many sustainable-use programs never move beyond the experimental stage or fall short of their objectives, they give an indication of some possible ways forward. The IUCN/SSC Sustainable Use Specialist Group involves wildlife managers and others all around the world who are actively engaged in efforts to understand how such schemes can work and why they don't work, and to share that understanding with local communities and conservation agencies seeking to enhance the prospects for wild species and wild lands that are under stress from human activities.

The development of "green" markets for sustainably produced goods and services is a growing sector that offers opportunities for both businesses and consumers to contribute to the conservation of wildlife and habitats. The range of consumer goods —from organic foods and environmentally friendly household cleaning agents to hybrid automobiles and furniture from recycled plastics— is already extensive and growing rapidly. "Green" certification schemes have been or are being developed to identify for consumers sustainably exploited timber and food and aquarium fishes, as well as hotels and other tourist destinations that actively contribute to conservation.

Challenging an ominous future

There are few who would not be discouraged by the statistics presented by the IUCN Red List: over 11 000 threatened plant and animal species, some 2 000 of them on the very brink of extinction, and thousands, possibly tens or hundreds of thousands, not yet assessed but believed to be at risk. And there are many who would find it difficult to see hope enough to engage in the challenge of reversing these trends. But whether or not one sees optimism in the many conservation achievements of the past several decades, the future of biodiversity and of our own existence —physical, material, and spiritual— depends on meeting that challenge.

Nature is the source of the raw materials and basic resources that we as humans need to survive. Every species that we lose forecloses future options to derive indirect or direct benefit from it in the future, such as through new medicines or materials, but also clean air to breathe and water to drink. Hence, extinction must be viewed not only for the tragedy that it is but as an omen of our life-support sys-

tem's breaking down, perhaps irretrievably. In destroying the species that link us to our evolutionary past, we also eliminate possibilities for our future.

Perhaps the greatest hope for averting future extinctions rests with a growing conservation constituency that individually and collectively is mobilizing efforts to conserve species and habitats and redress environmental degradation around the world. The IUCN Species Survival Commission, comprising over 7 000 scientists, conservation practitioners, and government officials from at least 188 countries, is not only part of that growing constituency but is fostering the collaboration that is so often necessary to achieve effective conservation action. These collaborative efforts, drawing on the skills, experience, and ingenuity of the best minds in conservation today, offer hope that we can meet the challenge of containing the sixth extinction wave. But that hope will only be realized through a universal, global commitment to change and to investing the resources to make that change.

Facing the extinction crisis and what it means for the future of life on Earth is not the sole responsibility of governments, scientists, the SSC, or other conservation bodies. It requires the engagement of all sectors of society and all players within those sectors around a fundamental ethical shift: not to shortchange the future in profiting from the present. We must renounce the common perception of nature as being disjunct from ourselves and in conflict with our aspirations and recognize that it is, instead, integral to these. And, in our individual actions and in the actions that we take and promote as corporate stockholders, constituents of government decision-makers, or members of other communities, we must specifically reduce the threats and redress the degradation that have brought so many species to the edge of extinction and ensure that others do the same. Although our knowledge is incomplete, it is more than we need to begin immediately to change course to making the right decisions for the future.

The next ten years will be crucial for many of the threatened species on the IUCN Red List. It may be even more so for the millions of species that have yet to be discovered. Many will be saved through the concerted efforts of individuals, institutions, and governments, but many will disappear, testimony of our inability to make fundamental changes in the way we as individuals, as nations, and as a world community, live our lives. The responsibility is ours. There is no time to lose.

FURTHER READINGS

Bight, Chris. 1998. *Life Out of Bounds. Bioinvasion in a Borderless World.* WorldWatch Institute. W.W. Norton & Co., New York.

Fitter, Richard and Maisie Fitter (eds.). 1987. *The Road to Extinction.* IUCN, Gland, Switzerland and Cambridge, U.K.

Lawton, John H. and Robert M. May (eds.). 1995. *Extinction Rates.* Oxford University Press, Oxford.

Stevens, William K. 1999. *The Change in the Weather. People, Weather, and the Science of Climate.* Dell Publishing, Random House, New York.

World Conservation Monitoring Centre. 2000. *Global Biodiversity. Earth's Living Resources in the 21st Century.* By B. Groombridge and M.D. Jenkins, World Conservation Press, Cambridge, U.K.

Selected Publications of IUCN and the IUCN/ SSC

Available from the IUCN Publications Services Unit, 219c Huntingdon Road, Cambridge CB3 0DL, United Kingdom or in North America from Island Press, Box 7, Covelo, California 95428, USA or through the on-line IUCN Bookstore at www.iucn.org.

IUCN and Related Red Lists of Threatened Species

2000 IUCN Red List of Threatened Species. 2000. Compiled by C. Hilton-Taylor. IUCN, Gland, Switzerland and Cambridge, U.K.

Threatened Birds of the World. 2000. BirdLife International. Lynx Edicions, Barcelona, and BirdLife International, Cambridge, U.K.

The World List of Threatened Trees. 1998. Compiled by S. Oldfield, C. Lusty, and A. MacKinven. World Conservation Press, Cambridge, U.K.

The 1997 IUCN Red List of Threatened Plants. 1998. Compiled by the World Conservation Monitoring Centre, Cambridge, U.K. Edited by K.S. Walter and H.J. Gillett.

Publications of the IUCN/SSC

Selected SSC Action Plans (arranged alphabetically by common name of species)

African Primates: Status Survey and Conservation Action Plan (revised edition). 1996. Compiled by John F. Oates and the IUCN/SSC Primate Specialist Group.

African Rhino: Status Survey and Action Plan. 1999. Compiled by Richard Emslie and Martin Brooks, and the IUCN/SSC African Rhino Specialist Group.

The African Wild Dog: Status Survey and Conservation Action Plan. 1997. Compiled and edited by Rosie Woodroffe, Joshua Ginsberg, David MacDonald, and the IUCN/SSC Canid Specialist Group.

Antelopes. Global Survey and Regional Action Plans. Part 3. West and Central Africa. 1990. Compiled by R. East and the IUCN/SSC Antelope Specialist Group.

The Asian Elephant. An Action Plan for Its Conservation. 1990. Compiled by C. Santiapillai, P. Jackson, and the IUCN/SSC Asian Elephant Specialist Group.

Asian Rhinos. Status Survey and Conservation Action Plan. Second edition. 1997. Edited by Thomas J. Foose, Nico van Strien, and the IUCN/SSC Asian Rhino Specialist Group.

Australasian Marsupials and Monotremes. An Action Plan for Their Conservation. 1992. Compiled by M. Kennedy and the IUCN/SSC Australasian Marsupial and Monotreme Specialist Group.

The 1996 Action Plan for Australasian Marsupials and Monotremes. 1996. Edited by S. Maxwell, A.A. Burbidge, K.D. Morris, and the IUCN/SSC Australasian Marsupial and Monotreme Specialist Group. Published by Wildlife Australia, Endangered Species Program.

Bears: Status Survey and Conservation Action Plan. 1998. Compiled by C. Servheen, H. Herrero, B. Peyton, and the IUCN/SSC Bear and Polar Bear Specialist Groups.

Cactus and Succulent Plants: Status Survey and Conservation Action Plan. 1997. Compiled by Sara Oldfield and the IUCN/SSC Cactus and Succulent Specialist Group.

Conifers: Status Survey and Conservation Action Plan. 1999. Compiled by A. Farjon, C.N. Page, and the IUCN/SSC Conifer Specialist Group.

The Cranes: Status Survey and Conservation Action Plan. 1996. Compiled by Curt D. Meine, George W. Archibald, and the IUCN/SSC Crane Specialist Group.

Crocodiles. Status Survey and Conservation Action Plan. Second edition. 1998. Edited by James Perran Ross. IUCN/SSC Crocodile Specialist Group.

Curassows, Guans, and Chachalacas: Status Survey and Conservation Action Plan for Cracids 2000-2004. 2000. Compiled by Daniel M. Brooks and Stuart D. Strahl, with Spanish and Portuguese translations.

Deer: Status Survey and Conservation Action Plan. 1998. Compiled by C. Wemmer and the IUCN/SSC Deer Specialist Group.

Dolphins, Porpoises, and Whales. 1994-1998 Action Plan for the Conservation of Cetaceans. 1994. Compiled by Randall R. Reeves and Stephen Leatherwood, together with the IUCN/SSC Cetacean Specialist Group.

Dragonflies: Status Survey and Conservation Action Plan. 1997. Compiled by Norman W. Moore and the IUCN/SSC Odonata Specialist Group.

The Ethiopian Wolf: Status Survey and Conservation Action Plan. 1997. Compiled and edited by Claudio Sillero-Zubiri, David MacDonald, and the IUCN/SSC Canid Specialist Group.

Eurasian Insectivores and Tree Shrews: Status Survey and Conservation Action Plan. 1996. Compiled by David Stone and the IUCN/SSC Insectivore, Tree Shrew, and Elephant Shrew Specialist Group.

Foxes, Wolves, Jackals, and Dogs. An Action Plan for the Conservation of Canids. 1990. Compiled by J.R. Ginsberg, D.W. MacDonald, and the IUCN/SSC Canid and Wolf Specialist Groups.

Grebes: Status Survey and Conservation Action Plan. 1997. Compiled by Colin O'Donnel, Jon Fjeldså, and the IUCN/SSC Grebe Specialist Group.

Grouse: Staus Survey and Conservation Action Plan 2000-2004. 2000. Compiled by Ilse Storch and the WPA/BirdLife/SSC Grouse Specialist Group.

Hyaenas: Status Survey and Conservation Action Plan. 1998. Compiled by Gus Mills, Heribert Hofer, and the IUCN/SSC Hyaena Specialist Group.

Insectivora and Elephant Shrews. An Action Plan for Their Conservation. 1990. Compiled by M.E. Nicoll, G.B. Rathbun, and the IUCN/SSC Insectivore, Tree Shrew, and Elephant Shrew Specialist Group.

Conservation of Mediterranean Island Plants. 1. Strategy for Action. 1996. Compiled by O. Delanoë, B. de Montmollin, L. Olivier, and the IUCN/SSC Mediterranean Islands Plant Specialist Group.

Megapodes: Status Survey and Conservation Action Plan 2000-2004. 2000. Edited by René W.R.J. Dekker, Richard A. Fuller, and Gillian C. Baker on behalf of the WPA/BirdLife/SSC Megapode Specialist Group.

Mosses, Liverworts, and Hornworts: Status Survey and Conservation Action Plan for Bryophytes. 2000. Compiled by Tomas Hallingbäck and Nick Hodgetts. IUCN/SSC Bryophyte Specialist Group.

North American Rodents: Status Survey and Conservation Action Plan. 1998. Compiled and edited by David J. Hafner, Eric Yensen, Gordon L. Kirkland Jr., and the IUCN/SSC Rodent Specialist Group.

Orchids: Status Survey and Conservation Action Plan. 1996. Edited by Eric Hágsater and Vinciane

Dumont, compiled by Alec Pridgeon and the IUCN/SSC Orchid Specialist Group.

Otters. An Action Plan for Their Conservation. 1990. Compiled by P. Foster-Turley, S. MacDonald, C. Mason, and the IUCN/SSC Otter Specialist Group.

Palms: Their Conservation and Sustained Utilization. Status Survey and Conservation Action Plan. 1996. Edited by Dennis Johnson and the IUCN/SSC Palm Specialist Group.

Parrots: Status Survey and Conservation Action Plan. 2000. Edited by Noel Snyder, Philip McGowan, James Gilardi, and Alejandro Grajal.

Partridges, Quails, Francolins, Snowcocks, Guineafowl, and Turkeys: Status Survey and Conservation Action Plan 2000-2004. 2000. Edited by Richard A. Fuller, John P. Carroll, and Philip J.K. McGowan on behalf of the WPA/BirdLife/SSC Partridge, Quail, and Francolin Specialist Group.

Pheasants: Status Survey and Conservation Action Plan 2000-2004. 2000. Edited by Richard A. Fuller and Peter J. Garson on behalf of the WPA/BirdLife/SSC Pheasant Specialist Group.

Pigs, Peccaries, and Hippos. Status Survey and Conservation Action Plan. 1993. Edited by William L.R. Oliver, the IUCN/SSC Pigs and Peccaries Specialist Group, and the IUCN/SSC Hippo Specialist Group.

Rabbits, Hares, and Pikas. Status Survey and Conservation Action Plan. 1990. Compiled by J.A. Chapman, J.E.C. Flux, and the IUCN/SSC Lagomorph Specialist Group.

The Red Panda, Olingos, Coatis, Raccoons, and Their Relatives. Status Survey and Conservation Action Plan for Procyonids and Ailurids. 1994. Compiled by Angela R. Glatston and the IUCN/SSC Mustelid, Viverrid, and Procyonid Specialist Group.

Seals, Fur Seals, Sea Lions, and Walrus. Status Survey and Conservation Action Plan. 1993. Compiled by Peter Reijnders, Sophie Brasseur, Jaap van der Toorn, Peter van der Wolf, Ian Boyd, John Harwood, David Lavigne, Lloyd Lowry, and the IUCN/SSC Seal Specialist Group.

South American Camelids. An Action Plan for Their Conservation. 1992. Compiled by H. Torres and the IUCN/ SSC South American Camelid Specialist Group.

Swallowtail Butterflies. An Action Plan for Their Conservation. 1991. Compiled by T.R. New, N.M. Collins, and the IUCN/SSC Lepidoptera Specialist Group.

Tapirs. Status Survey and Conservation Action Plan. 1997. Compiled by Daniel M. Brooks, Richard E. Bodmer, Sharon Matola, and the IUCN/SSC Tapir Specialist Group.

Tortoises and Freshwater Turtles. An Action Plan for Their Conservation. 1989. Compiled by the IUCN/SSC Tortoise and Freshwater Turtle Specialist Group.

West Indian Iguanas: Status Survey and Conservation Action Plan. 2000. Compiled and edited by Allison Alberts and the IUCN/SSC West Indian Iguana Specialist Group.

Wild Cats: Status Survey and Conservation Action Plan. 1996. Compiled and edited by Kristin Nowell, Peter Jackson, and the IUCN/SSC Cat Specialist Group.

Wild Sheep and Goats and Their Relatives. Status Survey and Conservation Action Plan for Caprinae. 1997. Edited by D.M. Shackleton and the IUCN/SSC Caprinae Specialist Group.

Zebras, Asses, and Horses. An Action Plan for the Conservation of Wild Equids. 1992. Compiled by P. Duncan and the IUCN/SSC Equid Specialist Group.

SSC Occasional Papers

Biology and Conservation of Asian River Cetaceans. 2000. Randall R. Reeves, B.D. Smith, and T. Kasuya. No. 23.

African Elephant Database 1998. 1999. R.F.W. Barnes, G.C. Craig, H.T. Dublin, G. Overton, W. Simmons, and C.R. Thouless. No. 22.

African Antelope Database 1998. 1999. Rod East. No. 21.

Sharks and Their Relatives: Ecology and Conservation. 1998. Merry Camhi, Sarah Fowler, John Musick, Amie Bräutigam, and Sonja Fordham. No. 20.

Proceedings of the Twelfth Working Meeting of the IUCN/SSC Polar Bear Specialist Group, 3-7 February 1997, Oslo, Norway. 1998. Compiled and edited by Andrew E. Derocher, Gerald W. Garner, Nickolas J. Lunn, and Oystein Wiig. No. 19.

Manejo y uso sustentable de pecaries en la Amazonia peruana. 1997. Richard Bodmer, Rolando Aquino, Pablo Puertas, César Reyes, Tula Fang, and Nicole Gottdenker. No. 18.

Sturgeon Stocks and Caviar Trade. Proceedings of a workshop held on 9-10 October 1995 in Bonn, Germany by the Federal Ministry for the Environment, Nature Conservation, and Nuclear Safety and the Federal Agency for Nature Conservation. 1997. Edited by Vadim J. Birstein, Andreas Bauer, and Astrid Kaiser-Pohlmann. No. 17.

The Live Bird Trade in Tanzania. 1996. Edited by N. Leader-Williams and R.K. Tibanyenda. No. 16.

Community-based Conservation in Tanzania. 1996. Edited by N. Leader-Williams, J.A. Kayera, and G.L. Overton. No. 15.

Tourist Hunting in Tanzania. 1996. Edited by N. Leader-Williams, J.A. Kayera, and G.L. Overton. No. 14.

Técnicas para el manejo del guanaco. 1995. Edited by Sylvia Puig, Chair of the South American Camelid Specialist Group. No. 13.

Assessing the Sustainability of Uses of Wild Species: Case Studies and Initial Assessment Procedure. 1996. Edited by Robert and Christine Prescott-Allen. No. 12.

African Elephant Database 1995. 1995. M.Y. Said, R.N. Chunge, G.C. Craig, C.R. Thouless, R.F.W. Barnes, and H.T. Dublin. No. 11.

Polar Bears: Proceedings of the Eleventh Working Meeting of the IUCN/SSC Polar Bear Specialist Group, January 25-28 1993, Copenhagen, Denmark. 1995. Compiled and edited by Oystein Wiig, Erik W. Born, and Gerald W. Garner. No. 10.

The Conservation Biology of Molluscs. Proceedings of a Symposium held at the 9th International Malacological Congress, Edinburgh, Scotland, 1986. 1995. Edited by Alison Kay. No. 9.

Conservation Biology of Lycaenidae (Butterflies). 1993. Edited by T.R. New. No. 8. (Out of print).

Polar Bears: Proceedings of the Tenth Working Meeting of the IUCN/SSC Polar Bear Specialist Group. 1991. No. 7.

Biodiversity in Sub-Saharan Africa and Its Islands: Conservation, Management, and Sustainable Use. 1991. Compiled by Simon N. Stuart and Richard J. Adams, with a contribution from Martin D. Jenkins. No. 6.

The Conservation Biology of Tortoises. 1989. Edited by I.R. Swingland and M.W. Klemens. No. 5. (Out of print).

Rodents. A World Survey of Species of Conservation Concern. 1989. Edited by W.Z. Lidicker, Jr. No. 4.

Priorités en matière de conservation des espèces à Madagascar. 1987. Edited by R.A. Mittermeier, L.H. Rakotovao, V. Randrianasolo, E.J. Sterling, and D. Devitre. No. 2. (Out of print).

Species Conservation Priorities in the Tropical Forests of Southeast Asia. 1985. Edited by R.A. Mittermeier and W.R. Konstant. No. 1. (Out of print).

IUCN Guidelines and Policy Statements on Species Issues

Guidelines for the Prevention of Biodiversity Loss Caused by Alien Invasive Species. Approved by the 51st Meeting of the IUCN Council, February 2000.

Guidelines for the Placement of Confiscated Animals. Approved by the 51st Meeting of the IUCN Council, February 2000.

The IUCN Red List Categories: Version 3.1. Approved by the 51st Meeting of the IUCN Council, February 2000.

Guidelines for Re-Introductions. Approved by the 41st Meeting of the IUCN Council, May 1995.

The IUCN Categories of Threat. Approved by the 40th Meeting of the IUCN Council, November 1994.

IUCN Policy Statement on State Gifts of Animals. Approved by the 27th Meeting of the IUCN Council, June 1989.

IUCN Policy Statement on Research Involving Species at Risk of Extinction. Approved by the 27th Meeting of the IUCN Council, June 1989.

The IUCN Position Statement on Translocation of Living Organisms: Introductions, Reintroductions and Re-Stocking. Approved by the 22nd Meeting of the IUCN Council, September 1987.

The IUCN Policy Statement on Captive Breeding. Approved by the 22nd Meeting of the IUCN Council, September 1987.

CONTRIBUTORS

DR. ALLISON ALBERTS
Zoological Society of San Diego
U.S.A.

DR. GIOVANNI AMORI
Consiglio Nazionale delle Ricerche
Italy

MR. BERTRAND VON ARX
Canadian Wildlife Service
Canada

DR. STANISLAV E. BELIKOV
All-Russian Research Institute for Nature
Russian Federation

DR. DANIEL M. BROOKS
Houston Museum of Natural Sciences
U.S.A.

DR. P. MARTIN BROOKS
KwaZulu-Natal Nature Conservation
Services
South Africa

DR. PHILLIP CRIBB
Royal Botanic Gardens, Kew
United Kingdom

DR. INDRANEIL DAS
Universiti Malaysia Sarawak (UNIMAS)
Malaysia

DR. PETER PAUL VAN DIJK
TRAFFIC Southeast Asia
Malaysia

DR. TERRY DONALDSON
International Marinelife Alliance
U.S.A.

DR. JOHN DRANSFIELD
Royal Botanic Gardens, Kew
United Kingdom

DR. M. EHSAN DULLOO
International Plant Genetic Resources
Institute
Kenya

DR. ROD EAST
NIWA Ecosystems
New Zealand

DR. PETER GARSON
University of Newcastle
United Kingdom

MR. SIMON HEDGES
United Kingdom

DR. MÓNICA HERZIG
FAUNAM, A.C.
Mexico

DR. CRAIG HILTON-TAYLOR
IUCN
United Kingdom

DR. BAZ HUGHES
Wildfowl and Wetlands Trust
United Kingdom

MR. ANTHONY M. HUTSON
United Kingdom

MR. PETER JACKSON
Switzerland

DR. WILLIAM KONSTANT
Conservation International
U.S.A.

DR. GERALD KUCHLING
University of Western Australia
Australia

DR. DAVID MACDONALD
Oxford University
United Kingdom

MS. SHARON MATOLA
Belize Zoo
Belize

MR. PHILIP MCGOWAN
World Pheasant Association
United Kingdom

DR. L. DAVID MECH
U.S. Geological Survey
U.S.A.

DR. GUS L. MILLS
National Parks Board
South Africa

DR. PATRICIA D. MOEHLMAN
Tanzania

DR. RICARDO OJEDA
CONICET-IADIZA
Argentina

MR. DAN L. PERLMAN
U.S.A.

MS. CAROLINE POLLOCK
IUCN
United Kingdom

DR. GALEN RATHBUN
California Academy of Sciences
U.S.A.

DR. CHRISTOPHER J. RAXWORTHY
American Museum of Natural History
U.S.A.

DR. RANDALL R. REEVES
Canada

DR. CLAUS REUTHER
Aktion Fischotterschutz e.V.
Germany

DR. YVONNE J. SADOVY
University of Hong Kong
Hong Kong

DR. SCOTT SCHLIEBE
U.S. Fish and Wildlife Service
U.S.A.

MR. CHRIS SHEPPARD
TRAFFIC Southeast Asia
Malaysia

DR. ANDREW SMITH
Arizona State University
U.S.A.

DR. ALISON STATTERSFIELD
BirdLife International
United Kingdom

DR. ILSE STORCH
Munich Wildlife Society
Germany

DR. WENDY STRAHM
IUCN
Switzerland

DR. WOLFGANG STUPPY
Royal Botanic Gardens, Kew
United Kingdom

DR. RAMAN SUKUMAR
Indian Institute of Science
India

DR. NIGEL TAYLOR
Royal Botanic Gardens, Kew
United Kingdom

DR. JORGEN THOMSEN
Conservation International
U.S.A.

THE 1994 IUCN RED LIST CATEGORIES

EXTINCT (EX)

A taxon is Extinct when there is no reasonable doubt that the last individual has died.

EXTINCT IN THE WILD (EW)

A taxon is Extinct in the Wild when it is known only to survive in cultivation, in captivity or as a naturalized population (or populations) well outside the past range. A taxon is presumed extinct in the wild when exhaustive surveys in known and/or expected habitat, at appropriate times (diurnal, seasonal, annual) throughout its historic range have failed to record an individual. Surveys should be over a time frame appropriate to the taxon's life cycle and life-form.

CRITICALLY ENDANGERED (CR)

A taxon is Critically Endangered when it is facing an extremely high risk of extinction in the wild in the immediate future.

ENDANGERED (EN)

A taxon is Endangered when it is not Critically Endangered, but is facing a very high risk of extinction in the wild in the near future.

VULNERABLE (VU)

A taxon is Vulnerable when it is not Critically Endangered or Endangered, but is facing a high risk of extinction in the wild in the medium-term future.

LOWER RISK (LR)

A taxon is Lower Risk when it has been evaluated and does not satisfy the criteria for any of the categories Critically Endangered, Endangered or Vulnerable. Taxa included in the Lower Risk category can be separated into three subcategories.

1. **Conservation Dependent (cd)**. Taxa which are the focus of a continuing taxon-specific or habitat-specific conservation program targeted towards the taxon in question, the cessation of which would result in the taxon qualifying for one of the threatened categories above within a period of five years.

2. **Near Threatened (nt)**. Taxa which do not qualify for Conservation Dependent, but which are close to qualifying for Vulnerable.

3. **Least Concern (lc)**. Taxa which do not qualify for Conservation Dependent or Near Threatened.

DATA DEFICIENT (DD)

A taxon is Data Deficient when there is inadequate information to make a direct, or indirect, assessment of its risk of extinction based on its distribution and/or population status. A taxon in this category may be well studied, and its biology well known, but appropriate data on abundance and/or distribution is lacking. Data Deficient is therefore not a category of threat or Lower Risk. Listing of taxa in this category indicates that more information is required and acknowledges the possibility that future research will show that threatened classification is appropriate. It is important to make positive use of whatever data are available. In many cases, great care should be exercised in choosing between DD and threatened status. If the range of a taxon is suspected to be relatively circumscribed and/or if a considerable period of time has elapsed since the last record of the taxon, threatened status may well be justified.

NOT EVALUATED (NE)

A taxon is Not Evaluated when it has not yet been assessed against the criteria.

TABLE 1. Rate of species description for major species groups

Group	Approximate number of new species recognized per year in the 1980s	Percentage annual increase in number of described species
Vertebrates	370	0.8
Birds	5	0.05
Mammals	30	0.6
Amphibians and reptiles	100	1.2
Fishes	230	1.2
Molluscs	370	0.5
Sponges	50	0.6
Cnidarians	60	0.6
Platyhelminths	320	1.6
Annelids	170	1.2
Protozoans	360	0.9
Crustaceans	700	1.7
Insects	7 200	0.8
Lepidopterans	640	0.4
Coleopterans	2 300	0.6
Dipterans	1 000	0.9
Hymenopterans	1 200	0.9
Arachnids	1 400	1.8
Nematodes	360	2.4
Fungi	1 700	2.4

NOTE: 171 mammals described between 1981 and 1992, including 86 rodents, 28 bats, and 23 insectivores. An additional 287 species recognized through taxonomic changes.

TABLE 2. Estimated number of described species vs. predicted total for major species groups

Kingdom	Group	Described species	Estimated total
Protoctists		80 000	600 000
Animals	Vertebrates	52 000	55 000
	Mammals	4 630	
	Birds	9 946	
	Reptiles	7 400	
	Amphibians	4 950	
	Fishes	25 000	
	Insects, centipedes, and millipedes	963 000	8 000 000
	Arachnids	75 000	700 000
	Molluscs	70 000	200 000
	Crustaceans	40 000	150 000
	Nematodes	25 000	400 000
Fungi		72 000	1 500 000
Plants		270 000	320 000

TABLE 3. Estimated number of described marine species

Kingdom	Group	Estimated marine species
Protoctists	Seaweeds (green, brown, and red algae)	12 500
	Diatoms, foraminiferans, and others	23 000
Animals	Porifera (sponges)	10 000
	Cnidaria (corals, jellyfishes, sea anemones)	10 000
	Flatworms	15 000
	Nematodes	12 000
	Bryozoans	5 000
	Molluscs	25 000
	Annelids	12 000
	Crustaceans	38 000
	Echinoderms	7 000
	Vertebrates	15 000
	Lampreys and hagfishes	50
	Sharks, skates, and rays	800
	Bony fishes	14 000
	Reptiles	60
	Birds	300
	Mammals	100
Fungi		500
Plants		50

TABLE 4. Estimated number of described inland water species

Kingdom	Selected group	Estimat. inland water spp.
Protoctists	Stoneworts and other large algae	500–600
Fungi		600
Plants	Ferns	250
	Seed plants	2 500
Animals	Molluscs	5 000
	Insects	45 000
	Ephemeroptera (mayflies)	2 250
	Odonata (dragonflies)	4 875
	Plecoptera (stoneflies)	2 100
	Hemiptera (bugs)	3 200
	Coleoptera (beetles)	5 000
	Diptera (flies)	over 20 000
	Trichoptera (caddisflies)	7 000
	Vertebrates	
	Fishes	10 000
	Amphibians	5 000
	Reptiles	220
	Birds	250
	Mammals	70

SOURCE FOR ALL TABLES: *2000 IUCN Red List of Threatened Species*. See previous page for abbreviations used in some of the tables.

TABLE 5. Recorded number of extinctions by major species group and biome according to 2000 IUCN Red List of Threatened Species

	EX			EW		
	Terrestrial	Inland Water	Marine	Terrestrial	Inland Water	Marine
Vertebrates						
Mammals	81	0	2	4	0	0
Birds	125	18	11	3	0	0
Reptiles	21	1	0	1	0	0
Amphibians	5	5	0	0	0	0
Fishes	0	81	1	0	11	0
Invertebrates						
Insects	64	14	0	1	0	0
Molluscs	199	88	4	9	3	0
Crustaceans	0	8	0	0	1	0
Others	4	0	0	0	0	0
Plants						
Mosses	3	0	0	0	0	0
Gymnosperms	0	0	0	1	0	0
Dicotyledons	69	0	0	14	0	0
Monocotyledons	1	0	0	2	0	0

NOTE: All species have been assigned to one or more biome; for example, seals are both marine and terrestrial; otters and amphibians are inland water and terrestrial; diadromous fishes are inland water and marine, etc. Inland waters include saline water bodies, cave waters, fresh waters, etc.

TABLE 6. Recorded extinctions by major species group according to 2000 IUCN Red List of Threatened Species

	EX	EW	Total
Vertebrates			
Mammals	83	4	87
Birds	128	3	131
Reptiles	21	1	22
Amphibians	5	0	5
Fishes	81	11	92
Subtotal	318	19	337
Invertebrates			
Insects	72	1	73
Molluscs	291	12	303
Crustaceans	8	1	9
Others	4	0	4
Subtotal	375	14	389
Plants			
Mosses	3	0	3
Gymnosperms	0	1	1
Dicotyledons	69	14	83
Monocotyledons	1	2	3
Subtotal	73	17	90
Total	766	50	816

NOTE: The counts are of species known to have become globally Extinct since AD 1500, although some of the mammals included are under dispute as they are allegedly only known from fossil evidence. There is no strictly comparable baseline for the plants, as the extinction criteria were different for the *1997 IUCN Red List of Threatened Plants*, and the figures now include non-trees, so they can not be compared to *The World List of Threatened Trees* (1998). The extinction data are in the process of being updated.

TABLE 7. Numbers of Red List threatened species in three major biomes

	Biome type		
	Marine	Inland water	Terrestrial
Mammals	25	31	1 111
Birds	105	78	1 144
Reptiles	9	111	283
Amphibians	0	131	143
Fishes	163	627	0
Spiders and centipedes	0	0	11
Crustaceans	0	409	0
Insects	0	125	438
Molluscs	13	420	508
Anthozoa	2	0	0
Enopla	2	0	0
Onychophora	0	0	6
Annelids	1	0	5
Plants	0	14	5 607
Total	320	1 946	9 256

NOTE: The counts are only for globally threatened species (CR, EN, VU). All species have been assigned to one or more biome. The molluscs were assigned with the help of Mary Seddon of the SSC Mollusc Specialist Group. Inland waters include saline water bodies, cave waters, fresh waters, etc. The plants, because they largely comprise trees, have been classified mainly as terrestrial, but some of the mosses included grow partially submerged in freshwater streams.

TABLE 8. 2000 IUCN Red List: Number of species assigned to each category by major biome

	Terrestrial	Inland water	Marine
EX	571	216	18
EW	35	15	0
CR	1 602	384	44
EN	2 199	503	67
VU	5 455	1 058	209
LR/cd	325	31	24
LR/nt	2 313	312	92
DD	1 157	453	182

NOTE: Species have been assigned to one or more biome; for example, seals are both marine and terrestrial; otters and amphibians are inland water and terrestrial; diadromous fishes (e.g., sturgeon) are inland water and marine, etc. Inland waters include saline water bodies, cave waters, fresh waters, etc.

Table 9. 2000 IUCN Red List status category summary by major taxonomic group (animals)

Class*	% of group assessed	EX	EW	Subtotal	CR	EN	VU	Subtotal	LR/cd	LR/nt	DD	Total
Mammalia	100	83	4	87	180	340	610	1 130	74	602	240	2 133
Aves	100	128	3	131	182	321	680	1 183	3	727	79	2 123
Reptilia	<15	21	1	22	56	79	161	296	3	74	59	454
Amphibia	<15	5	0	5	25	38	83	146	2	25	53	231
Cephalaspidomorphi	<10	1	0	1	0	1	2	3	0	5	3	12
Elasmobranchii		0	0	0	3	17	19	39	4	35	17	95
Actinopterygii		80	11	91	152	126	431	709	12	96	251	1 159
Sarcopterygii		0	0	0	1	0	0	1	0	0	0	1
Insecta	<0.1	72	1	73	45	118	392	555	3	76	40	747
Bivalvia	<5	31	0	31	52	28	12	92	5	60	7	195
Gastropoda		260	12	272	170	209	467	846	14	177	513	1 822
Crustacea	<5	8	1	9	56	72	280	408	9	1	32	459
Echinoidea	<0.1	0	0	0	0	0	0	0	0	1	0	1
Arachnida		0	0	0	0	1	9	10	0	1	7	18
Chilopoda		0	0	0	0	0	1	1	0	0	0	1
Merostomata		0	0	0	0	0	0	0	0	1	3	4
Onychophora		3	0	3	1	3	2	6	0	1	1	11
Hirudinoidea		0	0	0	0	0	0	0	0	1	0	1
Oligochaeta		0	0	0	1	0	4	5	0	1	0	6
Polychaeta		0	0	0	1	0	0	1	0	0	1	2
Enopla		0	0	0	0	0	2	2	0	1	3	6
Turbellaria		1	0	1	0	0	0	0	0	0	0	1
Anthozoa		0	0	0	0	0	2	2	0	0	1	3
Total		693	33	726	925	1 353	3 157	5 435	129	1 885	1 310	9 485

* Mammalia (mammals), Aves (birds), Reptilia (reptiles), Amphibia (amphibians), Cephalaspidomorphi (lampreys and hagfishes), Elasmobranchii (sharks, skates, rays, and chimaeras), Actinopterygii (bony fishes), Sarcopterygii (coelacanths), Insecta (insects), Bivalvia (mussels and clams), Gastropoda (snails, etc.), Crustacea (crustaceans), Echinoidea (sea urchins, starfishes, etc.), Arachnida (spiders and scorpions), Chilopoda (centipedes), Merostomata (horseshoe crabs), Onychophora (velvet worms), Hirudinoidea (leeches), Oligochaeta (earthworms), Polychaeta (marine bristle worms), Enopla (nemertine worms), Turbellaria (flatworms), Anthozoa (sea anemones and corals).

REFERENCE: World Conservation Monitoring Centre. 2000. *Global Biodiversity: Earth's Living Resources in the 21st Century.* By: B. Groombridge and M.D. Jenkins, World Conservation Press, Cambridge, U.K.

Table 10. 2000 IUCN Red List status category summary by major taxonomic group (plants)

Class*	% of group assessed	EX	EW	Subtotal	CR	EN	VU	Subtotal	LR/cd	LR/nt	DD	Total
Bryopsida	<1	2	0	2	10	15	11	36	0	0	0	38
Anthocerotopsida		0	0	0	0	1	1	2	0	0	0	2
Marchantiopsida		1	0	1	12	16	14	42	0	0	0	43
Coniferopsida	72	0	1	1	17	40	83	140	24	52	33	250
Ginkgoopsida		0	0	0	0	1	0	1	0	0	0	1
Magnoliopsida	<5	69	14	83	896	1 110	3 093	5 099	203	610	298	6 293
Liliopsida	<4	1	2	3	79	83	129	291	17	45	39	395
Total		73	17	90	1 014	1 266	3 331	5 611	244	707	370	7 022

* Bryopsida (mosses), Anthocerotopsida (hornworts), Marchantiopsida (liverworts), Coniferopsida (conifers), Ginkgoopsida (ginkgo), Magnoliopsida (dicotyledons), Liliopsida (monocotyledons). EX – Extinct, EW – Extinct in the Wild, CR – Critically Endangered, EN – Endangered, VU – Vulnerable, LR/cd – Lower Risk/conservation dependent, LR/nt – Lower Risk/near threatened, DD – Data Deficient.

NOTE: The proportions of species assessed are: mosses <1%; gymnosperms 72%; dicotyledons <5%; and monocotyledons <4%. These proportions are coarse approximations based on figures from WCMC (2000) and corrected using new information from the SSC's threatened species database or from the SSC Specialist Groups. The figures do not include any assessments from the 1997 plants Red List (Walter and Gillett 1998), as these were all done using the pre-1994 IUCN system of threat categorization. Hence, the very low proportions of plants assessed compared to the 1997 results. Similarly, the results can not be compared to *The World List of Threatened Trees* (1998), as other plants are now included.

REFERENCES: World Conservation Monitoring Centre. 2000. *Global Biodiversity: Earth's Living Resources in the 21st Century*. By: B. Groombridge and M.D. Jenkins, World Conservation Press, Cambridge, U.K. K.S. Walter and H. Gillett (eds.) 1998. *1997 IUCN Red List of Threatened Plants*. Compiled by the World Conservation Monitoring Centre. IUCN, Gland, Switzerland and Cambridge, U.K.

Table 11. Numbers of Red List threatened species by major groups of organisms

	Number of species in group	Number of threatened species in 2000	% of total in group threatened in 2000	% of total assessed threatened in 2000*
Vertebrates				
Mammals	4 763	1 130	24	24
Birds	9 946	1 183	12	12
Reptiles	7 970	296	4	25
Amphibians	4 950	146	3	20
Fishes	25 000	752	3	30
Subtotal	51 926	3 507	7	19
Invertebrates				
Insects	950 000	555	0.06	58
Molluscs	70 000	938	1	27
Crustaceans	40 000	408	1	20
Others	130 200	27	0.02	0.2
Subtotal	1 190 200	1 928	0.2	29
Plants				
Mosses	15 000	80	0.5	53
Gymnosperms	876	141	16	22
Dicotyledons	194 000	5 099	3	53
Monocotyledons	56 000	291	0.5	26
Subtotal	265 876	5 611	2	48

* NOTE: Threatened includes those listed as Critically Endangered (CR), Endangered (EN), and Vulnerable (VU). Other than mammals and birds, only a small or extremely small proportion of the total number of species in any group have been assessed for threatened status. The proportions of species assessed are: birds and mammals 100%; reptiles <15%; amphibians <15%; fishes <10%; insects <0.1%; molluscs <5%; crustaceans <5%; other invertebrates <0.1%; mosses <1%; gymnosperms 72%; dicotyledons <5%; and monocotyledons <4%. These proportions are coarse approximations based on figures from WCMC (2000) and corrected using new information from the SSC's threatened species database or from the SSC Specialist Groups. The plant figures do not include any assessments from the *1997 IUCN Red List of Threatened Plants*, as these were all done using the pre-1994 IUCN system of threat categorization. Hence, the very low proportions of plants assessed compared to the 1997 results. Similarly, the results can not be compared to *The World List of Threatened Trees* (1998), as other plants are now included.

SOURCES: Species numbers are mostly from WCMC (2000), except for mammals, where we have used and updated the data compiled by Mace and Balmford (2000), and reptiles, where we used the recent figures from the EMBL Reptile Database compiled by Peter Uetz (http://www.embl-heidelberg.de/~uetz/LivingReptiles.html), while plant numbers are from Farjon (1998), Hallingbäck and Hodgetts (2000), and Mabberley (1997), as corrected by Rudolph Schmid in *Taxon* 47:245 (1998).

IUCN SPECIES SURVIVAL COMMISSION DIRECTORY

Conservation Professionals Collaborating to Save the World's Biodiversity
The Species Survival Commission (SSC) is one of six volunteer commissions of IUCN-The World Conservation Union, a union of sovereign states, government agencies, and nongovernmental organizations. IUCN has three basic conservation objectives: to secure the conservation of nature, and especially of biological diversity, as an essential foundation for the future; to ensure that, where the Earth's natural resources are used, this is done in a wise, equitable, and sustainable way; and to guide the development of human communities towards ways of life that are both of good quality and in enduring harmony with other components of the biosphere.

The SSC's mission is to conserve biological diversity by developing and executing programs to save, restore, and wisely manage species and their habitats. A volunteer network comprising nearly 7 000 scientists, field researchers, government officials, and conservation leaders from nearly every country of the world, the SSC membership is an unmatched source of information about biological diversity and its conservation. SSC members provide technical and scientific counsel for conservation projects throughout the world and serve as resources to governments, international conventions, and conservation organizations.

Dr. David Brackett, Chair
Websites: IUCN Species Survival Commission
http://www.iucn.org/themes/ssc/index.htm
IUCN Red List of Threatened Species http://www.redlist.org
Newsletter: *Species*

The IUCN Species Survival Commission network is managed in collaboration with several other institutions, including the following:

BirdLife International. For over 20 years, BirdLife International has published information on globally threatened bird species, in regional Red Data Books (Africa in 1985, Americas in 1992, Asia in 2001) and global avian red lists (1988, 1994, and 2000). In terms of conservation status, birds are recognized as the best documented group of all species. BirdLife is the Listing Authority for birds for the IUCN Red List and, through its partnership of NGOs and Secretariat, works closely with the IUCN/SSC Specialist Groups and a worldwide network of thousands of experts and other organizations in this capacity. BirdLife makes its information on globally threatened birds available on the World Wide Web, supported by a dynamic database to manage, analyze, and report on these data. This wealth of information lays the foundation for BirdLife's work and guides its priorities for action.

Website: http://www.wing-wbsj.or.jp/birdlife/

Wetlands International. This is the world's leading wetland conservation organization, with offices in 16 countries and a coordinated network of thousands of experts around the globe, which include a number of Specialist Groups that form part of the IUCN Species Survival Commission. Its mission is to sustain and restore wetlands, their resources, and biodiversity for future generations through research, information exchange, and conservation activities worldwide. Wetlands International's approach to promoting wetlands conservation is based on acquir-

ing high-quality scientific information and providing this to governments, agencies, and organizations.

Website: http://www.wetlands.org

The World Pheasant Association (WPA). This is the world's foremost conservation body dedicated to ensuring the survival of the world's 300 species of galliform birds, pheasants, and pheasant relatives —grouse, partridges, cracids, and megapodes— and their habitats. Five galliform Specialist Groups operating under the auspices of WPA form part of the IUCN Species Survival Commission.

Website: http://www.pheasant.org.uk

IUCN/SSC Specialist Groups

MAMMALS

African Elephant Specialist Group
Dr. Holly T. Dublin, Chair
Newsletter: *Pachyderm*

African Rhino Specialist Group
Dr. P. Martin Brooks, Chair
Website: http://www.rhinos-irf.org/specialists
Newsletter: *Pachyderm*

Afrotheria Specialist Group
Dr. Galen Rathbun, Chair

Antelope Specialist Group
Dr. Richard D. Estes and Dr. Rod East, Cochairs
Newsletter: *Gnusletter*

Asian Elephant Specialist Group
Dr. Raman Sukumar, Chair
Newsletter: *GAJAH*

Asian Rhino Specialist Group
Dr. Mohammed Khan bin Momin Khan, Chair
Website: http://www.rhinos-irf.org/specialists
Newsletter: *Pachyderm*

Asian Wild Cattle Specialist Group
Dr. Simon Hedges, Chair
Newsletter: *Newsletter of the Asian Wild Cattle Specialist Group*

Australasian Marsupial and Monotreme Specialist Group
Dr. John H. Seebeck and Dr. Christopher J. Dickman, Cochairs

Bear Specialist Group
Newsletter: *International Bear News*

Bison Specialist Group
Dr. Cormack Gates and Dr. Wanda Olech, Cochairs

Canid Specialist Group
Dr. David MacDonald, Chair
Website: http://www.canids.org
Newsletter: *Canid News*

Caprinae Specialist Group
    Dr. Marco Festa-Bianchet, Chair
    Website: http://callisto.si.usherb.ca:8080/caprinae/iucnwork.htm
    Newsletter: *Caprinae News*

Cat Specialist Group
    Drs. Urs and Christine Breitenmoser, Cochairs
    Website: http://lynx.uio.no/catfolk
    Newsletter: *Cat News*

Cetacean Specialist Group
    Dr. Randall R. Reeves, Chair

Chiroptera Specialist Group
    Dr. Paul Racey and Mr. Anthony M. Hutson, Cochairs
    Newsletter: *Chiroptera Neotropical*

Deer Specialist Group
    Dr. Susana González, Chair
    Newsletter: *Newsletter of the Deer Specialist Group*

Edentate Specialist Group
    Dr. Gustavo Boucharde da Fonseca, Chair
    Newsletter: *Edentata*

Equid Specialist Group
    Dr. Patricia D. Moehlman, Chair
    Newsletter: *Equid Specialist Group Newsletter*

Hyaena Specialist Group
    Dr. Gus L. Mills, Chair

Insectivore Specialist Group
    Chair: Dr. Rainer Hutterer
    Website: http://members.vienna.at/shrew/itses.html
    Newsletter: *ITSES*

Lagomorph Specialist Group
    Dr. Andrew Smith, Chair
    Website: http://www.ualberta.ca/~dhik/lsg/

Mustelid, Viverrid, and Procyonid Specialist Group
    Mr. Roland Wirth, Chair
    Newsletter: *Small Carnivore Conservation*

New World Marsupial Specialist Group
    Dr. Ricardo Ojeda, Chair
    Website: http://www.cricyt.edu.ar/institutos/iadiza/ojeda/marsupiales.htm

Otter Specialist Group
    Dr. Claus Reuther, Chair
    Newsletter: *IUCN/SSC Otter Specialist Group Bulletin; IUCN/SSC Asian Otter*

Pangolin Specialist Group
    Dr. Jung-Tai Chao, Chair

Pig, Peccary, and Hippo Specialist Group
    Mr. William Oliver, Chair

Polar Bear Specialist Group
    Dr. Stanislav E. Belikov and Dr. Scott Schliebe, Cochairs

Primate Specialist Group
    Dr. Russell Mittermeier, Chair
    Newsletters: *Primate Conservation, Lemur News, Neotropical Primates, African Primates, Asian Primates*

Rodent Specialist Group
    Dr. Giovanni Amori, Chair

Seal Specialist Group

Sirenia Specialist Group
    Dr. John Reynolds and Dr. James Powell, Cochairs
    Newsletter: *Sirenews*

South American Camelid Specialist Group
    Dr. Silvia Puig, Chair
    Newsletter: *Cartas Noticias*

Tapir Specialist Group
    Dr. Patricia Medici, Chair
    Website: http://www.tapirback.com/tapirgal/iucn-ssc/tsg/

Wolf Specialist Group
    Dr. L. David Mech, Chair
    Website: http://home.kassel.netsurf.de/oliver.matla/ewn_e.htm
    Newsletter: *European Wolf Newsletter*

BIRDS

Bustard Specialist Group
    Dr. Asad Rafi Rahmani, Chair

Cormorant Specialist Group
    Dr. Mennobart Van Eerden, Chair
    Newsletter: *Wetlands International Cormorant Research Group Bulletin*

Cracid Specialist Group
    Dr. Daniel M. Brooks, Chair
    Website: http://www.angelfire.com/ca6/cracid/
    Newsletter: *Bulletin of the Cracid Specialist Group*

Crane Specialist Group
    Dr. George Archibald, Chair
    Website: http://www.savingcranes.org
    Newsletter: *ICF Bugle*

Diver and Loon Specialist Group
    Dr. Joseph Kerekes, Chair
    Website: http://www.briloon.org/Diver.html
    Newsletter: *Diver/Loon Specialist Group Newsletter*

Duck Specialist Group
    Dr. Jeff Kirby, Chair

Flamingo Specialist Group
    Dr. Alan Johnson, Chair
    Website: http://www.wetlands.org/sgroups/flamingo/flamingo.htm
    Newsletter: *Flamingo Research*

Goose Specialist Group
    Dr. Bart Ebbinge, Chair
    Website: http://www.wetlands.org/sgroups/goose/goose.htm
    Newsletter: *Wetlands International Goose Specialist Group Bulletin*

Grebe Specialist Group
    Dr. Jon Fjeldså, Chair

Grouse Specialist Group
    Dr. Ilse Storch, Chair
    Website: http://www.pheasant.org.uk/gsg.asp

Heron Specialist Group
    Dr. Heinz Hafner, Chair
    Website: http://www.wetlands.org/sgroups/heron/heron.htm
    Newsletter: *Heron Conservation Newsletter*

Megapode Specialist Group
Dr. Rene W.R.J. Dekker, Chair
Website: http://www.ens.gu.edu.au/ecology/darryl/megap.htm
Newsletter: *Megapode Newsletter*

Partridge, Quail, and Francolin Specialist Group
Dr. John P. Carroll, Chair
Website: http://www.game-conservancy.org.uk/pqf/
Newsletter: *Newsletter of the Partridge, Quail, and Francolin Specialist Group*

Pelican Specialist Group
Dr. Alan Crivelli, Chair

Pheasant Specialist Group
Dr. Peter Garson, Chair
Website: http://www.pheasant.org.uk/psg.asp
Newsletter: *Tragopan*

Rail Specialist Group
Dr. Barry Taylor, Chair

Seaduck Specialist Group
Dr. Stefan Pihl, Chair
Website: http://www.dmu.dk/coastalzoneecology/seaduck/index.html
Newsletter: *Seaduck Bulletin*

Stork, Ibis, and Spoonbill Specialist Group
Dr. Malcolm Coulter, Chair
Newsletter: *Specialist Group on Storks, Ibises, and Spoonbills Newsletter*

Swan Specialist Group
Dr. Jan Beekman, Chair
Website: http://www.wetlands.org/sgroups/swan/swan.htm
Newsletter: *Swan Specialist Group Newsletter*

Threatened Waterfowl Specialist Group
Dr. Baz Hughes, Chair
Website: http://www.wwt.org.uk/threatsp/twsg/
Newsletter: *TWSG Bulletin*

Wader Specialist Group
Mr. David Stroud, Chair
Website: http://www.wetlands.org/sgroups/wader/wader.htm

Woodcock and Snipe Specialist Group
Dr. Heribert Kalchreuter, Chair
Website: http://www.wetlands.org/sgroups/wood_snipe/wood_snipe.htm
Newsletter: *Woodcock and Snipe Specialist Group*

REPTILES AND AMPHIBIANS

African Reptiles Specialist Group
Dr. William Branch, Chair
Newsletter: *Newsletter of the IUCN/SSC African Reptile and Amphibian Specialist Group*

China Reptile and Amphibian Group
Prof. Li Pi Peng, Chair

Crocodile Specialist Group
Dr. Harry Messel, Chair
Website: http://www.flmnh.ufl.edu/natsci/herpetology/crocs/crocsd.htm
Newsletter: *IUCN/SSC Crocodile Specialist Group*

European Reptile and Amphibian Specialist Group
Mr. Keith F. Corbett, Chair

Global Amphibian Specialist Group
Dr. Claude Gascon, Chair

Iguana Specialist Group
Dr. Allison Alberts and Dr. Jose Ottenwalder, Cochairs
Website: http://www.scz.org/iguana
Newsletter: *West Indian Iguana Specialist Group Newsletter*

Madagascar and Mascarene Reptile and Amphibian Specialist Group
Dr. Gerald Kuchling, Chair
Newsletter: *Madagascar and Mascarene Reptiles and Amphibians*

Marine Turtle Specialist Group
Dr. Alberto Abreu Grobois, Chair
Newsletter: *Marine Turtle Specialist Group Bulletin*

South American Reptile Specialist Group
Dr. Jorge Williams, Chair
Website: http://www.netverk.com.ar/~jorgew/herpetolist/

South and Southeast Asian Reptile Specialist Group
Dr. Indraneil Das, Chair

Tortoise and Freshwater Turtle Specialist Group
Dr. John Behler and Dr. Anders Rhodin, Cochairs
Newsletter: *Chelonian Conservation and Biology*

FISHES

Caribbean Fish Specialist Group
Dr. Michael Smith and Dr. Carlos Rodríguez, Cochairs
Website: http://caribbeanfish.org/

Coral Reef Fish Specialist Group
Dr. Terry Donaldson, Chair

Grouper and Wrasse Specialist Group
Dr. Yvonne J. Sadovy, Chair
Website: http://www.hku.hk/ecology/GroupersWrasses/iucnsg/index.html

Shark Specialist Group
Ms. Sarah Fowler and Dr. John A. Musick, Cochairs
Website: http://www.flmnh.ufl.edu/fish/Organizations/SSG/
Newsletter: *Shark News*

Sturgeon Specialist Group
Dr. Mohammad Pourkazemi, Chair

INVERTEBRATES

Inland Water Crustacean Specialist Group
Dr. Keith Crandall, Chair
Newsletter: *Anostracan News*

Mollusc Specialist Group
Dr. Mary Seddon, Chair
Website: http://bama.ua.edu/~clydeard/IUCN-SSC_html/index.htm
Newsletter: *Tentacle*

Odonata Specialist Group
Dr. Jan Van Tol, Chair
Newsletter: *Reports of the Odonata Specialist Group*

Social Insect Specialist Group
Dr. Donat Agosti, Chair
Website: http://research.amnh.org/entomology/social_insects/sisg.html

Southern African Invertebrate Specialist Group
    Prof. Michael Samways, Chair

PLANTS

Arabian Plants Specialist Group
    Dr. Abdul-Aziz Abuzinada, Chair

Australasian Plants Specialist Group
    Ms. Jeanette Mill, Chair
    Website: http://www.anbg.gov.au/anpc/danhome.html
    Newsletter: *Newsletter of the Australian Network for Plant Conservation*

Bamboo Specialist Group

Bryophyte Specialist Group
    Dr. Tomas Hallingbäck, Chair
    Website: http://www.dha.slu.se/guest/SSCBryo/SSCBryo.html

Bulb Specialist Group
    Dr. Alan Meerow, Chair

Cactus and Succulent Specialist Group
    Website: http://wwwcjb.unige.ch/BVAUICN/BNL.htm
    Newsletter: *Cacti and Succulent Group Newsletter*

Carnivorous Plant Specialist Group
    Mr. Bertrand von Arx, Chair
    Newsletter: *Carnivorous Plants Specialist Group Newsletter*

China Plant Specialist Group
    Dr. Wang Xianpu and Dr. Qui Haining, Cochairs

Conifer Specialist Group
    Dr. Aljos Farjon, Chair
    Newsletter: *FITZROYA*

Cycad Specialist Group
    Dr. John Donaldson, Chair

Eastern Africa Plant Specialist Group
    Ms. Stella Simiyu, Chair

European Plant Specialist Group
    Dr. Klaus Ammann, Chair

Fungi Specialist Group
    Dr. Régis Courtecuisse, Chair
    Newsletter: *Fungi Conservation Newsletter*

Indian Ocean Island Plant Specialist Group
    Dr. M. Ehsan Dulloo and Dr. Dominique Strasberg, Cochairs

Indian Subcontinent Plant Specialist Group
    Dr. C.R. Babu, Chair

Japanese Plant Specialist Group
    Dr. Tetsukazu Yahara, Chair

Korean Plant Specialist Group
    Dr. Kim Yong-Shik, Chair

Lichen Specialist Group
    Dr. Christoph Scheidegger, Chair

Macronesian Island Plant Specialist Group
    Dr. Julia Pérez de Paz, Chair

Medicinal Plant Specialist Group
    Dr. Danna Leaman, Chair
    Website: http://www.dainet.de/genres/mpc-dir/
    Newsletter: *Medicinal Plant Conservation*

Mediterranean Island Plant Specialist Group
    Dr. Bertrand de Montmollin, Chair

North American Plant Specialist Group
    Ms. Peggy Olwell, Chair
    Website: http://www.nps.gov/plants/index.htm

Orchid Specialist Group
    Dr. Phillip Cribb, Chair
    Newsletter: *Orchid Conservation News*

Palm Specialist Group
    Dr. William Hahn, Chair

Philippine Plant Specialist Group
    Dr. Domingo Madulid, Chair
    Newsletter: *Philippine Plants Specialist Group Newsletter*

Pteridophyte Specialist Group
    Dr. Tom Ranker and Mr. Clive Jermy, Cochairs
    Newsletter: *Fern Conservation News*

South Atlantic Island Plant Specialist Group

Southern African Plant Specialist Group
    Dr. Chris Willis, Chair

Temperate Broadleaved Tree Specialist Group
    Newsletter: *Broadleaves*

Temperate South American Plant Specialist Group
    Dr. Carlos Villamil, Chair

DISCIPLINARY

Conservation Breeding Specialist Group
    Dr. Ulysses S. Seal, Chair
    Website: http://www.cbsg.org/
    Newsletters: *CBSG News, CBSG India News, CBSG News - Indonesia*

Invasive Species Specialist Group
    Dr. Mick Clout, Chair
    Website: http://www.issg.org/
    Newsletter: *Aliens*

Reintroduction Specialist Group
    Dr. Frederick Launay, Chair
    Newsletter: *Reintroduction News*

Sustainable Use Specialist Group
    Mr. Leif Christoffersen, Chair
    Website: http://www.iucn.org/themes/sui/publications.html
    Newsletter: *Sustainable*

Veterinary Specialist Group
    Dr. William Karesh and Dr. Richard Kock, Cochairs
    Newsletter: *Newsletter of the IUCN/SSC Veterinary Specialist Group*

IUCN/SSC Task Forces and Committees

Declining Amphibian Populations Task Force (DAPTF)
    Dr. Jim Hanken, Chair
    Website: http://www.open.ac.uk/daptf/
    Newsletter: *FROGLOG*

Plant Conservation Committee
    Dr. David Given, Chair
    Website: http://wwwcjb.unige.ch/BVAUICN/BPLANTS.HTM

Fossa *Cryptoprocta ferox* (Endangered)

The largest of Madagascar's eight endemic carnivores, the fossa (pronounced "foosh") is also one of the largest members of the family Viverridae. A solitary, nocturnal hunter with an undeserved reputation for ferocity, the fossa is widely distributed on the island, but has been affected by habitat destruction and persecution as a chicken-thief. © Pete Oxford/BBC Natural History Unit

Red Sea soft corals and fairy basslet

As is the case with much of the Indo-Pacific, the Red Sea harbors a marvelous array of marine life, including many species that are found nowhere else. It is particularly well known for the abundance and variety of its soft corals. However, like other tropical marine systems, it is very fragile and under threat from coastal development, oil drilling, shipping, and other human activities. © Mike Bacon

Lesser long-nosed bat *Leptonycteris curasoae* (Vulnerable)

The lesser long-nosed bat is a specialized feeder on nectar from the flowers of cacti and other succulents, which it in turn pollinates by transferring pollen trapped in its fur from one flower to another. Ranging from Arizona, U.S.A. into northern South America, it breeds in caves and disused mineshafts in colonies that may number many thousands of individuals and are very susceptible to disturbance. © Merlin D. Tuttle/Bat Conservation International

Tropical rainforest, Mexico

Mexico is one of the most biologically rich countries in the world. Formerly covering vast tracts of southeastern Mexico, the tropical rainforests of the country have almost disappeared: less than 10% still remains, in scattered patches, scarcely but sadly suggestive of their former glory. Over 60% of the plant species distributed in the floristic province of Yucatán are endemic to Mesoamerica. Known for their very high biodiversity and structural complexity, these tropical rainforest ecosystems are at serious risk of disappearing altogether in a few decades due to overexploitation and misuse. © Patricio Robles Gil/Sierra Madre

Jaguar *Panthera onca* (Lower Risk/near threatened)

Why is this animal like a magnet? What does this animal have that captivates us? As Patricio Robles Gil has stated: "I do not think there is another animal that can be as good an ambassador of the tropical rainforests of the Americas. It fulfills all the requirements: beauty, strength, cunning, speed, and above all, it is an enigmatic animal of which we only know a little." © Stafan Widstrand

Pilbara Karijini National Park, Australia

The second largest park in Western Australia and the traditional home of the Banyjima, Kurrama, and Innawonga peoples, this national park harbors a variety of animals and plants. The drama of the landscape is consistent with extremes of climate, including variable temperatures and rainfall, and cyclones. © Bill Bellson/Lochman Transparencies

Captions written by Amie Bräutigam, Ramón Pérez Gil, and Mónica Herzig, with the information provided by contributors.

Victoria crowned pigeon *Goura victoria* (Vulnerable)

This spectacular bird inhabits the lowland forests of northern New Guinea, where it feeds on the ground in small groups of 2-10 individuals and roosts in trees. It is not well known. Stunning in appearance, it is also very easy to hunt, and much sought-after for meat, as well as feathers, which have been exported to international markets in recent years. Hunting is known to have reduced populations, and logging and capture for trade are presumed to be additional threats. © Gerald Cubitt

Mountain gorilla *Gorilla beringei beringei* (Critically Endangered)

The mountain gorilla vies for the honor of being the world's largest living primate, yet its enormous size and strength belie this creature's gentle nature. Found only in the mist-enshrouded Virunga Mountains of eastern equatorial Africa, this magnificent great ape is also one of the world's most endangered species, numbering only a few hundred individuals. Its remaining populations are essentially trapped on an "island" of montane forest surrounded by a "sea" of agricultural land and human settlements. However, field research, tourism, and conservation efforts have helped ensure its survival, even through times of intense regional conflicts. © Stafan Widstrand

Desert tortoise *Gopherus agassizii* (Vulnerable)

This terrestrial reptile is found in the southwestern U.S. and in Northwest Mexico in isolated, localized populations, many of which have disappeared or been depleted as a result of hunting for food, habitat loss from agriculture, grazing by livestock, and other pressures. It digs deep, long burrows for shelter from predators and extreme temperatures. © Patricio Robles Gil/Sierra Madre

Asian elephant *Elephas maximus* (Endangered)

Occurring in thirteen countries, the Asian elephant numbers fewer than 45 000 animals in the wild and some 16 000 in captivity. Habitat loss and hunting for ivory and other products are major threats, and populations are also declining through control of elephant depredations of agricultural crops. According to Dr. Raman Sukumar, Chairman of the IUCN/SSC Asian Elephant Specialist Group, conservation projects that aim at maintaining relatively large, viable populations in intact landscapes of multiple use hold the only prospects for the long-term conservation of this species. © Patricio Robles Gil/Sierra Madre

Nilgiri tahr *Hemitragus hylocrius* (Endangered)

The Nilgiri tahr is a mountain goat once common in southern India but now found only in small numbers in isolated pockets in the Western Ghats region of the country. Quick and sharp-sighted, it travels in flocks of varying sizes. Habitat degradation and poaching are two factors indicated in the decline of this species, which numbers only some 2 000 animals. © Patricio Robles Gil/Sierra Madre

California condor *Gymnogyps californianus* (Critically Endangered)

As a carrion-feeder dependent upon the carcasses of large mammals for food, the California condor was probably never numerous in the wild. Persecution by humans, in the form of shooting and poisoning, at the turn of the twentieth century far exceeded this bird's naturally low reproductive rate. It disappeared, except from California, in 1937 and declined steadily until 1987 when, after a great deal of controversy, the last wild condor was taken into captivity to join 26 others. This successful captive-breeding and reintroduction program may turn the tide: in 1998, the total population reached 150 birds, and 35 of those were in the wild. © Art Wolfe

True toad *Bufo typhonius*

This rainforest species forms part of an oddly attractive group of amphibians characterized by their cryptic coloration and unusual physique —most individuals have pointed, leaf-like noses. Although widespread through the Amazon Basin, their long-term welfare is far from assured: the decline of amphibian populations is a global problem, attributable to many factors, worthy of being addressed as an environmental emergency. The SSC's Declining Amphibian Populations Task Force has documented this phenomenon and its wider implications. © Art Wolfe

Big-headed turtle *Platysternon megacephalum* (Endangered)

At least 100 species of freshwater turtles and tortoises occur in Asia, and at least a third of these are threatened with extinction. The threats faced by Asian turtles are as diverse as the species themselves, but collection for domestic and international trade, which is measured in the thousands of tons, is currently the greatest menace to the survival of most species. The big-headed turtle was once common in the markets of southern China but now is sold in very small numbers, indicating a severe decline in populations in the wild. © Peter Paul van Dijk

Bristlecone pine *Pinus longaeva* (Vulnerable)

A native of the U.S.A., where it is still widely distributed on higher mountain peaks in California, Nevada, and Utah, the bristlecone pine is thought to include some of the world's oldest living organisms. This remarkable tree is extremely slow-growing and long-lived. Many trees are known to be over 4 000 years old, and one individual has been aged at around 5 200 years. The study of growth rings in living and fallen dead trees has yielded much valuable information on climate change over the past 9 000 years, and also indicated that the species has retreated in range over the last several thousand years. Although it occurs in remote areas, most populations are dominated by very old or senescent trees, and it is doubtful that current regeneration rates are sufficient to replace the existing population. © Françoise Gohier/Ardea London Ltd.

Big-belly sea horses *Hippocampus abdominalis* (Vulnerable)

Reproduction in sea horses is remarkable in that the female lays her eggs in the male's pouch, and the male fertilizes and nourishes the eggs until he gives birth to 100-250 fully-formed young sea horses of about 1 cm in length, which then swim away to care for themselves. Most sea horse species mate for life. A huge international trade in sea horses for traditional medicines, aquarium pets, and curios is depleting species' numbers in many countries. © Rudie Kuiter/Innerspace Visions

Mountain pygmy possum *Burramys parvus* (Endangered)

The world's only hibernating marsupial, the mountain pygmy possum is endemic to Australia and found only in mountainous regions where there is a continuous period of snow cover for up to 6 months of the year. It is threatened by destruction of its habitat from skiing and other winter sport facilities, as well as predation by foxes and feral cats. It is also believed to face an additional threat from global warming, which could reduce the snow cover it requires for insulation during the winter. © Jean-Paul Ferrero/Auscape

Spotted handfish *Brachionichthys hirsutus* (Critically Endangered)

Named for its habit of "walking" on its pelvic and pectoral fins rather than swimming, the spotted handfish was one of the first Australian fish species known to science and the first marine fish to be listed as endangered under the Australian Endangered Species Act. It is found only in the lower Derwent River Estuary in Tasmania where, in addition to low fertility (it lays only 80-250 eggs) and dispersal ability, it is believed to be threatened by predation of its eggs by the introduced northern Pacific sea star. A recovery plan is being implemented through the collaboration of at least six governmental and nongovernmental agencies, and initial captive-breeding trials and other efforts in favor of this species have so far proved encouraging. © Bill & Peter Boyle/Auscape

Starfish (*Formia nodosa*)

Sea stars, or starfishes, are, in many ways, the quintessential marine animal. They may also be the most unusual of the sea's more well-known creatures: they have no front or back and move in any direction without turning. As many as 1 500-2 000 species occur in the world's oceans. Some have as many as 28 arms, others as few as five; even more strangely, certain species include individuals with different numbers of arms. Their conservation status is completely unknown. © Mike Bacon

Fitzgerald River Biosphere Reserve, Australia

Renowned for its spectacular topography and biodiversity, this national park protects at least 19 native mammals and many other threatened animals, more than 1,800 species of flowering plants, and a wide array of lichens, mosses, and fungi. Quite a number of these plants are unique to this area. The scarlet banksia is one of 73 banksia species, members of the Proteaceae family, of which 58 occur only in southwestern Australia. Their dense flower spikes and curious fruit make them a distinctive feature of these bushlands.© Frans Lanting

Baby aye-aye *Daubentonia madagascariensis* (Endangered)

Madagascar's aye-aye easily qualifies as the world's most unlikely primate. In addition to its ghoulish appearance, this nocturnal species bears certain physical traits not found in any other primates, including ever-growing incisor teeth similar to those of rodents and an elongated middle finger that it uses to scoop insect larvae from beneath tree bark and hard fruits. The aye-aye is considered a harbinger of bad luck throughout Madagascar and is often killed on sight by villagers. However, it still occurs in more than 20 national parks and reserves, where it enjoys some level of protection. © Konrad Wothe

Strangler fig *Ficus* spp.

The tropical rainforests of the world harbor a wide variety of "strangler fig" trees, members of the Moraceae family, which includes 40 genera and 1 000 species. Of these, 11% are currently recorded as threatened, a fourth of which belong to the genus *Ficus*. This group may actually be more threatened than is currently known, as pollination and seed dispersion in these species are closely related to wasps and bats, whose status in many instances is also of concern, in particular in areas such as Sulawesi that have been heavily deforested. These interconnections are seldom thought about, but are critical. © Michael & Patricia Fogden

Palm *Marojejya darianii* (Endangered)

Discovered in 1983 and known from a single population, this swamp-dwelling palm, endemic to Madagascar, has very restricted areas of distribution. With species like this, even minor local disturbances to the habitat can have critical results. Demand for agricultural land to feed a growing population, coupled with outdated production methods, have resulted in increases in forest clearance and habitat destruction. Intensive collecting by overzealous palm growers also represents a serious threat, as all seed is collected and so rejuvenation is arrested. © Andrew McRobb/RBG, Kew

Yabbie crayfish *Cherax destructor* (Vulnerable)

The southeastern United States and Victoria and Tasmania, Australia are the two centers of diversity for the world's 600 species of freshwater crayfishes. The scientific name of this Australian yabby (freshwater crayfish), destructor, refers to its habit of burrowing into and damaging levees and dam walls. This species makes for excellent eating and is raised widely in aquaculture operations in Australia and elsewhere. It is a menace when it invades suitable habitat. © Jiri Lochman/ Lochman Transparencies

Humphead wrasse *Cheilinus undulatus* (Vulnerable)

Widely distributed in the Indo-Pacific, this magnificent animal has long been considered a noteworthy fish in the Pacific, and is served during ceremonies or for special occasions. In recent decades, it has been captured in increasing numbers for a rapidly expanding live reef-fish food trade. The total wholesale value of live reef fish imported into Hong Kong, the demand center for this trade, was estimated in 1998 to exceed US$500 million per year. This species is known to have declined as a result of overfishing. © Doug Perrine/Innerspace Visions

Valle de los Cirios, Mexico

The harsh conditions of the desertic Baja California Peninsula in northwestern Mexico have favored the evolution of bizarre and unusual plants, among them the boojum tree or cirio *Idria columnaris*, the endemic elephant tree, torote or copalquín *Pachycormus discolor*, and the near-endemic cardon cactus *Pachycereus pringlei*, often mistaken for the saguaro *Carnegiea gigantea*. The boojum tree is the tallest plant in this desert. © Jack Dykinga

Blue bird of paradise *Paradisaea rudolphi* (Vulnerable)

New Guinea's birds of paradise are amongst the world's most distinct living organisms. Luxuriant plumage and elaborate courtship display are prominent characteristics of these polygamous, promiscuous species, which are descended from an ordinary raven-like bird that reached this island in prehistoric times. Loss of forest habitat and hunting for its pectoral and tail feathers, valued for ceremonial and decorative purposes, are the main threats to this species. © Silvestris Fotoservice/A.N.T.

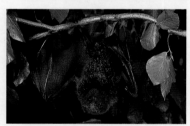

Rodrigues flying fox *Pteropus rodricensis* (Critically Endangered)

The Rodrigues flying fox is one of 160-odd species of fruit bat in the family Pteropodidae which play an important role as pollinators and seed dispersers in the Old World tropics. Like many other species in this group, it has a very limited distribution, having once occurred on both Rodrigues and the nearby island of Mauritius, from where it disappeared many years ago. The once-abundant Rodrigues population was reduced by the mid-1970s to fewer than 100 animals through a combination of habitat destruction, hunting, and severe cyclone devastation. Subsequent reforestation efforts and protection of the bats have increased numbers to 1 500-2 000, 80% of which are found at three roost sites in one forest area. © Michael Durham/gerryellis.com

Madagascar chameleon *Furcifer minor* (Vulnerable)

Madagascar is home to half of the world's chameleon species, including the largest and the smallest in this group of spectacular but little-known lizards. *Furcifer minor* is known only from the southern slopes of the island's central plateau. The larger adult males (up to 30 cm) are easily distinguished from the much smaller (only 15 cm long) and much more colorful females. © Christopher Raxworthy

Comet moth *Argema mittrei*

The butterflies and moths that make up the insect order Lepidoptera are some of the most familiar, spectacular, and appreciated of the world's insects. Diverse in number, form, and adaptation, they occur in a wide variety of habitats around the world. Although just under 300 species are listed as threatened or of concern in the IUCN Red List, this represents a tiny proportion of the over 100 000 species in this group that are likely to be at risk. © Art Wolfe

Giant anteater *Myrmecophaga tridactyla* (Vulnerable)

The giant anteater has a wide range in South America, where it occurs in swampy areas, humid forests, and savannahs. It eats up to 30 000 ants and termites in a single day. Docile and inoffensive by nature, its principal natural enemies are the puma and the jaguar. Its numbers have declined as a result of hunting for meat and trophies and loss of habitat. © Kevin Schafer

Monkey puzzle tree *Araucaria araucana* (Vulnerable)

The monkey puzzle tree is a native of southern South America, where it is found from the coastal cordillera of Chile to the Andes in Argentina. This is a region of intense volcanism, and this tree is well-adapted to fire, with thick, insulating bark and an ability to regenerate from burnt stumps. It is logged for timber, and its seeds are collected as an important source of food for humans and livestock. Populations on the coast are restricted and highly threatened, and those in the Andes are now severely fragmented. Despite legal protection, the species is still logged in both Argentina and Chile. © Alan Watson/Forest Light

Long-beaked echidna *Zaglossus bruijni* (Endangered) and Great spotted kiwi *Apteryx haastii* (Vulnerable)

Mammal or bird? Strikingly similar in appearance, the long-beaked echidna, a mammal, and the great spotted kiwi, a bird, are fundamentally different animals but resemble each other uncannily in a number of ways. The long-beaked echidna is one of only three species of egg-laying mammals (which also suckle their young) called monotremes. Found only on the island of New Guinea in mid-montane grassland and montane forest habitats, it is nocturnal and, with its long beak, forages in forest litter for the earthworms that make up the major part of its diet. Although still numerous, the long-beaked echidna is declining in numbers as a result of habitat loss and being hunted for food. Found in more or less the same corner of the globe, the great spotted kiwi is the largest of the kiwis, a group of wingless, flightless birds occurring only in New Zealand. This kiwi occurs only on South Island, where its range has shrunk and become severely fragmented since European settlement, and several populations have completely disappeared. An inhabitant of a variety of habitats on forested mountains, it, like the echidna, feeds primarily on invertebrates, but also consumes fruits and leaves. It is relatively long-lived and lays only one egg. Introduced predators including cats, dogs, and pigs are currently the greatest threat to this species. (Echidna) © D. Parer & E. Parer-Cook/Ardea London Ltd.; (kiwi) © Don Hadden/Ardea London Ltd.

King bird of paradise *Cicinnurus regius* (Lower Risk/least concern)

One of more than 40 bird-of-paradise species occurring in New Guinea and adjacent islands, this species is not considered at risk at this time. Birds of paradise are known for their bright plumage and elongated tail feathers, which are highly valued for ceremonial and decorative purposes. They are widely hunted for their feathers, and trade is known to occur, but actual numbers and what they may mean for the conservation of these species are unknown. © Brian D. Coates/Silvestris Fotoservice

Coral shrimp and bubble coral, Papua New Guinea

Anemone shrimp and bubble coral shrimps and other crustaceans are one of the most important groups of animals living on coral reefs. Many of these are "reef doctors" who clean parasites off from fishes or have other special, mutually beneficial relationships with corals, anemones, and other reef animals. Shrimps that live amongst the stinging tentacles of anemones can shelter themselves from predators while helping attract the anemone's next meal. © Mike Bacon

Baikal teal *Anas formosa* (Vulnerable)
In the early twentieth century, this was one of the most numerous ducks in eastern Asia, with flocks of several thousand birds regularly reported. However, since the 1960s and 70s, there has been a serious decrease in numbers. Hunting was probably the main reason for its decline and continues to be a serious threat, particularly as it concentrates in large flocks on arable land, but its wetland habitat is also being lost to agricultural and other development. © Konrad Wothe

Coconut or robber crab *Birgus latro* (Data Deficient)
The world's largest land invertebrate, the coconut crab reaches a leg span of over 1 m and a weight of 3-4 kg. Found throughout the western Pacific and eastern Indian Ocean, primarily on islands, it is highly valued for food and has great cultural importance. Overexploitation, in certain instances worsened by habitat loss, is known to have reduced populations in some areas, but in many its status and conservation needs are unknown. © Jean-Paul Ferrero/Auscape

Kagu *Rhynochetos jubatus* (Endangered)
The kagu is the only member of its family Rhynochetidae. This "ghostly-grey" flightless bird is endemic to the island of New Caledonia in the southwest Pacific, where its stronghold of some 300 birds is the Parc Provincial Rivière Bleue. It inhabits a variety of forest types, but is usually found in humid forest and sometimes extends into closed-canopy scrub during the wet season. It feeds on worms, snails, and lizards. Predation by dogs, introduced rats, and feral pigs is a threat, as is the degradation of its habitat through erosion from mining, fires, and logging. The kagu has been successfully bred in captivity since 1978 and reintroduced to protected areas. © Jean-Paul Ferrero/Auscape

Yellow-eyed penguin *Megadyptes antipodes* (Endangered)
During the breeding season, the yellow-eyed penguin occupies a tiny area in southern New Zealand, on South Island and outlying islands. The scrubland and forest habitats that it occupies for nesting have declined in quality, and it faces other threats on land and when it takes to the sea, including predation by introduced ferrets and stoats, as well as cats, and drowning in fishing nets. Periodic shortages in squid, sprats, and other prey caused by sea-temperature changes may also be a problem. The total world population of the yellow-eyed penguin is believed to number 4 000 birds, with varying trends in different areas. A wide range of conservation and research activities, which include fencing of nesting sites, trapping of predators, and habitat restoration, offer hope for the recovery of this species. © Kevin Schafer

Tuatara *Sphenodon guntheri* (Vulnerable)
Once regarded as a single species, the two species of tuatara are the only living members of the order Rhynchocephalia, a group of reptiles that were once common throughout the world 120-225 million years ago. Having become extinct on the main islands of New Zealand, they now occur only on 30 small, relatively inaccessible islands off the coast, *S. guntheri* being found on North Brother Island. © Kevin Schafer/Latinstock

Angelfishes and coral reefs, Thailand
Often referred to as the "rainforests of the sea," coral reefs are known for the great diversity of the species that they harbor. Some 4 000 fishes inhabit the world's coral reefs, and they vary greatly in form and shape in order to take advantage of the different types of habitats within these reefs. This "cluster" of angelfishes seems frozen in mid-action, as if expecting a hungry predator to appear suddenly to scatter them all over the reef. © 2000 Norbert Wu/www.norbertwu.com

Quiver trees *Aloe dichotoma* (Lower Risk/least concern)

One of the most attractive of the tree aloes of southern Africa, *Aloe dichotoma* is named for its use by the indigenous San people of the hollowed-out branches as quivers to carry their arrows. Found on rocky slopes in very dry regions of South Africa and Namibia, this spectacular plant grows, tree-like, to 4-5 m tall and 1 m in diameter, and flowers from May to July. The Kokerboom Forest Reserve was declared in 1993. © Michael & Patricia Fogden

Weevil beetle (family Curculionidae)

Like many of its kind, this tropical weevil beetle carries in its head the trademark of its feeding habits. Many species feed on the bountiful harvest of grains and seeds found in tropical rainforests such as Los Tuxtlas, and their striking mandibles represent a wonderful adaptation for boring into the particular kind of seed they feast upon. © Claudio Contreras Koob

Zanzibar red colobus *Procolobus kirkii* (Endangered)

The Zanzibar red colobus is a particularly attractive monkey, light-colored underneath and dark-red, brown, and black above, with distinct pink lips and nose on a black muzzle and a crown of long, white hairs around its face. It is found on the island of Unguja, off the coast of Zanzibar, where very little of the original tropical evergreen forest remains. The red colobus must survive in a mosaic of secondary forest, mangrove swamps, agricultural lands, and fallow bush. Fewer than 2 000 animals are thought to be left, but visitors can easily see the Zanzibar red colobus in the Jozani Forest Reserve. It is hoped that the benefits of ecotourism will help ensure its survival. © Patricio Robles Gil/Sierra Madre

Bongo *Tragelaphus eurycerus* (Lower Risk/near threatened)

One of the most majestic of the 70-odd antelope species of sub-Saharan Africa, the bongo inhabits the continent's equatorial forests where, according to Dr. Rod East, Cochairman of the IUCN/SSC Antelope Specialist Group, it is threatened by the habitat destruction and illegal meat-hunting that are accompanying the "relentless spread of human activities into Africa's remaining wilderness areas." © Anup Shah

Red-ruffed lemur *Varecia variegata rubra* (Critically Endangered)

This wonderfully attractive lemur is, like all lemurs, restricted to the island of Madagascar, where it inhabits the rainforests of the Masoala Peninsula. Forest clearance for agriculture has heavily degraded parts of this region, and hunting has also been a problem. A good number of animals in captivity may offer a hedge against its extinction. © Anup Shah

Fiji banded iguana *Brachylophus fasciatus* (Endangered)

This rare and beautiful lizard is found only on the South Pacific islands of Fiji and Tonga. Similarities to New World iguanas suggest that its ancestors may have crossed the ocean from South America (9 000 km away) relatively recently by drifting with the south equatorial currents on rafts of floating vegetation. © Thomas Wiewandt

Mandrill *Mandrillus sphinx* (Vulnerable)

Native to Western Equatorial Africa, this elusive species, an endemic forest-living relative of the savanna baboons, is strongly affected by human disturbance. Although it is not considered among the most threatened African primate species, in most areas drills are threatened by loss of habitat. Hunted for their meat, mandrills are especially vulnerable to hunters because of their loud calls. Currently, their densities are very low and they are poorly protected, if at all. © Kevin Schafer

Tessellated pupfish *Cualac tessellatus* (Endangered)

With a life span of less than one year, pupfish are tiny fish that live in springs, ponds, marshes, and slow-flowing streams in the deserts of southwestern North America, mainly in the Mojave and Sonoran deserts. Many types of endemic pupfish, such as this one, are endangered due to water pollution, modification and loss of habitat (groundwater pumping, stream channeling, water impoundment and diversion) and to competition from exotic species of fish that have been introduced to their habitat. © Pablo Cervantes/Sierra Madre

Flowering giant lobelia and greenswords, Hawaii

As is typical of islands, Hawaii is noted for the very high rate of endemism (87%) of its 1 048 vascular plant species; the rate of endemism of its plant genera is the highest in the world. One of the most spectacular groups of plants to have evolved in Hawaii is the lobelias. Many of Hawaii's plants are at risk: less than 40% of the land surface of the islands features native vegetation, and most of that is at higher elevations. © Frans Lanting

Marsh deer *Blastocerus dichotomus* (Vulnerable)

Although some deer species are abundant, many have fallen victim to the profound changes brought about by human activities. The marsh deer, the largest of the South American species, was formerly abundant across the seasonally flooded grasslands of south-central South America, but is declining throughout this range as a result of habitat loss and degradation and overhunting for meat, hide, and antlers. © Patricio Robles Gil/Sierra Madre

Malayan pangolin *Manis javanica* (Lower Risk/near threatened)

Cloaked in armor of overlapping scales and adapted to a specialized diet of ants and termites, pangolins are bizarre creatures. Being primarily nocturnal, they are also little known. All seven pangolin species (four in Africa and three in Asia) are widely used by humans for food and medicinal and ritualistic purposes. The Asian trade in pangolin scales, thought to focus primarily on the Malayan pangolin, has been measured in metric tons. Efforts to assess the importance of exploitation and trade for pangolins have been stymied by lack of knowledge of these species. © Alain Compost

Tarsier *Tarsius bancanus* (Data Deficient)

Best-known for its enormous glowing eyes and tiny size, the western tarsier is another of the world's unlikely primates. It is also amongst the most poorly known. Native to Southeast Asia, where it lives primarily in heavily-forested areas and is active at night, it feeds mostly on insects, but is also known to eat small animals, even an occasional poisonous snake. In addition to loss of habitat, tarsiers are heavily hunted for food and captured as pets, which is thought to be putting them at risk. © Frans Lanting

Long-nosed or proboscis monkey *Nasalis larvatus* (Endangered)

This large, long-tailed monkey with its reddish-brown and gray coat is known only from the island of Borneo in Southeast Asia. Its long, pendulous nose and honking call are trademarks of the adult male. Smaller-nosed females may weigh as much as 10 kg, but the males are often twice that size and sport prominent canine teeth which they use in threat displays. Proboscis monkeys live in palm, swamp, and mangrove forests, where they forage for a variety of leaves, seeds, fruits, and flowers. Concern for their survival has increased in recent years, as their numbers appear to have dwindled to fewer than 10 000 as a result of habitat loss and hunting. © Günter Ziesler

Hector's dolphin *Cephalorhynchus hectori* (Endangered)

Found only in the waters of New Zealand, Hector's dolphin is the rarest of the marine dolphins, with a world total of only 3 000-4 000 animals in three separate populations, one of which, the Critically Endangered North Island population, numbers fewer than 100 dolphins. The main threat is accidental drowning in fishing nets. Efforts are under way to establish no-fishing zones to protect this and other species from this fisheries bycatch. © Ingrid Visser/Innerspace Visions

Dragonfly (order Odonata)

Over 6 000 species of Odonata have been described worldwide, and more await description. The Odonata Specialist Group, the first organization established specifically to further the conservation of dragonflies, stated: "Since dragonflies have virtually no known economic importance, financial support for research is negligible." "There are huge gaps in our knowledge and slowness in acquiring it," while destruction of habitats continues. "In these circumstances, only a broad-brush approach can hope to conserve the hundreds of species under threat." © Claudio Contreras Koob

Tiger *Panthera tigris* (Endangered)

While poaching for bones and other parts, primarily for use in traditional Chinese medicine, is a better-known threat to the tiger, decline in prey is a more serious factor. Even if poaching were stopped altogether, the meat-eating tiger could still disappear from its wild habitats in Asia due to increasing threats to sambar deer and other favored prey. Three tiger subspecies are already extinct. © Patricio Robles Gil/Sierra Madre

Tupai *Tupaia longipes* on *Rafflesia arnoldi*

The species that comprise this group of Asian tree shrews owe their name "tupaia" to the Malay word for squirrel. The massive forest fires that have plagued Sumatra in recent years, coupled with illegal and unregulated logging, have undoubtedly affected these animals and their habitats, thus calling into question their long-term future. © Alain Compost

Bilby *Macrotis lagotis* (Vulnerable)

One of Australia's many endemic marsupials, the bilby inhabits the more arid regions of the continent. Its range and numbers have declined as a result of several factors including predation by introduced foxes, grazing by livestock and introduced rabbits, and changes in the fire regime, which is believed to be important in creating habitat and promoting the production of food resources. © Jiri Lochman/Lochman Transparencies

Komodo dragon *Varanus komodoensis* (Vulnerable)

Inhabiting a cluster of islands in the Lesser Sundas of Indonesia where, according to one early twentieth-century explorer, "every prospect pleases, and only man is vile," the Komodo dragon is the world's largest lizard, growing to over 3 m long and 100 kg in weight. Its serrated teeth, powerful claws, acute sense of smell, and sudden bursts of speed make it a formidable predator (and scavenger) of deer, wild boars, and other mammals —and other komodos. The total world population of about 5 000 komodos lives in Komodo Island National Park and on the island of Flores, where they face the combined threats of depletion of their prey from hunting and loss of habitat. Despite official protection, the future of the komodo, as prophesied decades ago, depends on curtailing destructive human activities. © Adrian Warren/Ardea London Ltd.

Galápagos giant tortoise *Geochelone nigra* (Vulnerable)

The Galápagos giant tortoise is the largest living tortoise, weighing over 225 kg and measuring 2 m from head to tail. Once numerous over the islands of the Galápagos Archipelago, this tortoise was heavily hunted by whalers and naval ships centuries ago and, since then, has been severely affected by introduced domestic pests and livestock. Very few young are being born into the different island populations, and captive-breeding/restocking and feral mammal control programs are having mixed success. © Stafan Widstrand

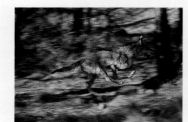

Mexican wolf *Canis lupus baileyi* (Extinct in the Wild)

The gray wolf has the largest natural range of any living terrestrial mammal other than *Homo sapiens*. However, it has disappeared from much of that range because of persecution by humans. Throughout the 1800s and into the 1900s, ranchers and farmers killed off the Mexican wolf until only a handful survived. The last wild wolves were brought into captivity for captive breeding, and in 1998, 11 were released into the wild in the Blue Range Wolf Recovery Area in Arizona. © Patricio Robles Gil/Sierra Madre

White uakari *Cacajao calvus calvus* (Endangered)

The white uakari is a little-known denizen of flooded white-water Amazonian forests. Its shaggy white coat, a rarity among the monkeys of the Americas, stands out in striking contrast to its fiery red face and balding head. Uakaris live in some of the largest groups ever recorded for Amazonian primates, sometimes numbering 100 or more animals. Their jaws and teeth are incredibly strong, adapted to cracking open fruits and nuts. They are threatened by restriction of their habitat and hunting. © Luiz Claudio Marigo

Leatherback turtle *Dermochelys coriacea* (Critically Endangered)

One of seven species of marine turtle widely distributed across the world's oceans, the leatherback has been listed as threatened for many years, but was recently reclassified as Critically Endangered as a result of a near-total collapse of the Pacific population of the species, which until recently was its global stronghold, and marked declines in the Atlantic population. The major cause of the leatherback's decline is incidental capture, or bycatch, in commercial fisheries, in particular longline fisheries. Other threats include overcollection of eggs from nesting beaches for human consumption. © Luiz Claudio Marigo

Asiatic lion *Panthera leo persica* (Critically Endangered)

The only living representatives of the lions once found throughout much of southwest Asia occur solely in India's Gir Forest, which was formerly the private hunting grounds of the Nawab of Junagadh. The total population numbers in the low hundreds and is believed to be rising, but the lions face many threats, including increasing conflicts with humans living in and around the sanctuary. © Anup Shah

Gharial *Gavialis gangeticus* (Endangered)

Named for the bulb or "ghara" (meaning "pot") at the tip of its long, thin snout, the gharial is one of the most endangered of the crocodilians. It occurs in very small numbers in several river systems in the northern Indian subcontinent, where it feeds almost exclusively on fish. Conservation programs, in particular captive breeding and restocking of wild populations, have been under way since the late 1970s and have brought it back from the brink of extinction. © Michael & Patricia Fogden

Giant river otter *Pteronura brasiliensis* (Endangered)

The giant river otter is the largest of the world's thirteen otter species and, arguably, the most charismatic. Once widespread throughout the rainforests, wetlands, and water bodies of South America, it was severely reduced by heavy hunting for pelts during the last two centuries and is now threatened by many pressures, including increased human colonization of tropical lowland forests, logging, and draining of wetlands, as well as water pollution from industrial and agricultural activities and mining operations. © André Bärtschi

West Indian manatee *Trichechus manatus* (Vulnerable)

Formerly widespread along the coasts of the Gulf of Mexico and the West Indies, this odd-looking "sea cow" is facing trouble due to a range of human activities including boating (in Florida, in particular), fishing, and coastal development; water pollution; hunting (in some areas); and the destruction of the seagrass beds on which it feeds. It is slow-moving and prefers shallow waters, which renders it particularly susceptible to these impacts. © Mike Bacon

Small giant clam *Tridacna maxima* (Lower Risk/conservation dependent)

One of nine species of giant clams inhabiting the warm waters of the tropical Indo-Pacific, *Tridacna maxima* is also the most widespread and still reasonably abundant. Giant clams have formed part of the diet of Pacific islanders and coastal dwellers in Australasia for thousands of years. They continue to be collected for their meat and shells and, more recently, as live animals for the aquarium trade. © Fred Bavendam/Minden Pictures

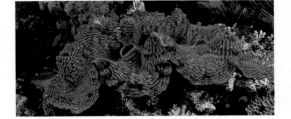

Ricord's iguana *Cyclura ricordi* (Critically Endangered)

The iguanas of the West Indies are among the largest and most impressive, but also are subject to the greatest threat. They live in one of the most endangered ecosystems on Earth: the dry tropical forests of the wider Caribbean. Ricord's iguana is found only in the southwestern Dominican Republic, and its population of some 2 000-4 000 animals is declining from habitat loss, predation by introduced cats, dogs, and mongooses, and hunting for food. © Doug Wechsler

Golden toad *Bufo periglenes* (Critically Endangered)

One of a number of amphibian species that have mysteriously disappeared over the past 15 years, the golden toad may be extinct: it has not been seen since 1989. An inhabitant of a small area of 4 km² in the Monteverde Cloud Forest Preserve in northern Costa Rica, the golden toad is only active above ground for a few days at certain times of the year, when it emerges to breed in temporary pools. Drier conditions caused by a decrease in cloud cover have recently been suggested as causing this species' disappearance. © Michael & Patricia Fogden

Great bustard *Otis tarda* (Vulnerable)
Among the largest of the flying birds, the great bustard, with its majestic bearing and elaborate mating dance, once ranged across the grasslands of Eurasia but has declined severely over several centuries, a victim of overhunting, urbanization, and industrialized agriculture. It now occurs in scattered remnant populations, and current trends in the destruction of its grassland habitat, in particular in eastern Europe, Russia, and central Asia, suggest that its long-term future is bleak. © Carlos Sánchez

Lesser prairie chicken *Tympanuchus pallidicinctus* (Vulnerable)
One of the relatively few bird species endemic to the United States, this animal once occurred in the vast grasslands of the southern Great Plains region, but is now restricted to small, scattered portions of this range. Well-known for its spectacular courtship displays, it has lost much of its habitat to industrialized agriculture and continues to be affected by habitat-related problems, including the effects of pesticides and herbicides. © Kenneth W. Frink/Ardea London Ltd.

Beetle *Chrysina* sp.
Encased in a shell of metallic beauty, this beetle stands poised on its leafy throne, seemingly unaware of the fate that may befall its species and that of a myriad of other beetle species all over the world, as their habitats continue to disappear at an increasingly alarming rate. Even if it were to number in the millions, that would likely not guarantee its survival. © Claudio Contreras Koob

Baird's tapir *Tapirus bairdii* (Vulnerable)
The largest forest-dwelling mammal in Mesoamerica and one of four surviving species of tapirs, Baird's tapir is closely associated with marshes and other humid habitats. Like their cousins, these animals are important seed-dispersers. Habitat loss and hunting, as well as large individual home ranges and a slow reproductive rate, are factors in their decline: they are considered endangered in most countries where they occur. © Patricio Robles Gil/Sierra Madre

Fernandina's flicker *Colaptes fernandinae* (Endangered)
The Zapata Swamp in Cuba harbors 60% of an estimated world population of just 100 pairs of this endemic flicker. This sociable woodpecker still thrives in the locality due to the wealth of sabal palms, now under threat from conversion to agriculture and hunting for food. This palm, the keystone species of this ecosystem, provides shelter and food for a wide array of species. If something happens to the palm, the entire ecosystem may be at risk. © Doug Wechsler/Vireo

Cycad *Encephalartos hildebrandtii*
Named for the nineteenth-century German botanist and explorer who first collected it, this attractive cycad is found in the wild in seasonally dry woodlands near the coasts of Kenya and Tanzania. In times of food shortages, it is used to make "ugali," a type of porridge. A popular ornamental plant, it is widely cultivated. © Andrew McRobb/RBG, Kew

Pearson's tree frog *Litoria pearsoniana* (Vulnerable)

Tree frogs are but one group of dozens of Australian frog species that seem to be disappearing right before our eyes. Frog declines appeared to have started in Australia in the late 1970s and are still continuing: more than 30 species have become extinct over the past 15 years. While the causes of decline of localized populations can be readily attributed to a given factor such as urbanization, invasive fishes, and so on, the reasons for most frog declines are still unclear. © Hans & Judy Beste/Lochman Transparencies

Beluga or great sturgeon *Huso huso* (Endangered)

The source of beluga caviar, one of the world's most coveted commodities, the beluga sturgeon is also the largest freshwater fish: in the past, individuals reached a length of *ca.* 6 m and a weight of 1 500 kg. An inhabitant of the Caspian, Black, and Adriatic Seas and the rivers flowing into them, it has declined from overfishing and the construction of dams on the Volga, Don, and Danube rivers. Natural reproduction is extremely limited, and some populations are entirely dependent on the release of juveniles from commercial sturgeon farms. Rampant poaching and illegal trade in caviar may wipe out this species. © P. Morris/Ardea London Ltd.

Great curassow *Crax rubra* (Lower Risk/near threatened)

One of 50 species in the family Cracidae, the most threatened group of birds in the Americas, the great curassow inhabits cloud forests from Mexico through to Ecuador where, with its cousins, it is an important seed-disperser and source of food for humans. Widespread destruction of tropical forests and overhunting have seriously affected these magnificent birds and put many species at serious risk of extinction. © Fulvio Eccardi

Green peafowl *Pavo muticus* (Vulnerable)

This majestic animal, native to Indo-China, has undergone a rapid population decline, in part due to high hunting levels, but also to habitat decline. It survives in a few scattered remnant populations in a variety of habitats in China and half a dozen other countries. The male's spectacular train feathers are traded in significant numbers. Regarded as a crop pest by farmers in China, it is often poisoned. © Patricio Robles Gil/Sierra Madre

Knowlton cactus *Pediocactus knowltonii*

Only four populations of this dwarf cactus have been recorded since its discovery in 1958. Two of these are now gone and a third has recently been almost totally destroyed. Due to private and commercial collecting, it is now one of the rarest cactus species in the United States, with only several hundred plants found in the wild. Although threatened by habitat disturbances (gas drilling, recreation, and livestock), collecting pressures have declined and some recovery has recently been observed. However, at the present population level, it is easily conceivable that the act of one irresponsible collector could eliminate the species. © George H.H. Huey

Great white shark *Carcharodon carcharias* (Vulnerable)

The world's best-known —and possibly most heavily-persecuted— fish, the great white shark, is also a species at risk. Although it is widely distributed across the world's oceans, it occurs in very limited numbers and has low reproductive output, which make it particularly vulnerable to overfishing. It is not well known, but in some areas has clearly declined in numbers. It may be more threatened than we know. © James D. Watt/Innerspace Visions

Staghorn beetle (family Cerambycidae)

Beetles are the most numerous group of animals on Earth. Over 350 000 beetle species have been described so far, and there may be as many as 3 million living in nearly every habitat except the open sea. In addition, beetles include some of the largest of the world's insects. Some stag beetles are known to be threatened, but there has been no systematic review of their status; there is little doubt that more are threatened than we now know. © Mark Moffett/Minden Pictures

Philippine eagle *Pithecophaga jefferyi* (Critically Endangered)

This magnificent bird has been holding on for several decades to ever-dwindling forest habitat in the Philippines. It faces numerous threats and is estimated to number as few as 350-650 birds. However, various conservation initiatives have been launched, which include protective legislation, nest-protection, field surveys, public awareness campaigns, and captive breeding. A pilot project designed to ease pressure on eagle territories while providing opportunities for economic development for local communities in these areas has had promising results. The future of the Philippine eagle will only be secured through a full implementation of a forest-conservation program that will also benefit 27 other threatened forest bird species occurring in the same areas. © Günter Ziesler

Cape fynbos, South Africa

"Fynbos" (fine bush) is the collective name for the main vegetation type that characterizes the Cape Floral Kingdom of South Africa. Fynbos embraces, for the most part, low-growing, small-leafed, drought-resistant, evergreen shrubs, many of them endemic and vulnerable to extinction. Especially varied is the plant family Proteaceae (about 370 species), which includes: the king protea *Protyea cynaroides*, South Africa's national flower; the sugar-bush *Protea repens*; and the lovely silver tree *Leucadendron argenteum*, which shimmers in the Cape breezes. © Colin Paterson-Jones

Boto dolphin *Inia geoffrensis* (Vulnerable)

The boto is the least threatened of the four strictly freshwater river dolphins. It occurs throughout most of the Amazon and Orinoco river drainages of South America, where populations appear to be large and there is little or no evidence of any decrease in numbers or range. As human activities increase and intensify, however, this situation is expected to change. The most significant current impact on the boto appears to be mortality in fishing operations, either incidental (entanglement), or to some extent deliberate: fishermen are reportedly poisoning botos with baited fish to limit net depredation and also shooting and otherwise killing animals found in or near to nets. © Flip Nicklin/Minden Pictures

Mediterranean monk seal *Monachus monachus* (Critically Endangered)

The most endangered of the world's seal species, now found in tiny pockets in the eastern Mediterranean and off the northwest African coast, the Mediterranean monk seal was dealt a terrible blow in 1997, when a massive die-off hit the northwest African population. Two thirds of the colony died, and only just over 100 animals survived. A toxic algal bloom is thought to have been the cause. © Francisco Márquez

Pygmy hippopotamus *Hexaprotodon liberiensis* (Vulnerable)

At first dismissed as a stunted freak, a dwarf subspecies or a juvenile specimen of the common hippopotamus, the pygmy hippo is more pig-like than its larger relative. A solitary creature, less aquatic than the common hippo, it inhabits dense forests along rivers and swamps, mainly in Liberia, in western Africa. It is not well known. Deforestation is considered to be the main threat to this species, although hunting for meat is also thought to put pressure on wild populations. © Anup Shah

Wandering albatross *Diomedea exulans* (Vulnerable)

Perhaps best-known for its being cursed by sailors as a harbinger of bad luck, this bird of the world's southern oceans spends long periods at sea, covering several thousand kilometers on a single foraging trip, and comes to shore seasonally to nest. It is long-lived, with an estimated life-span of 30-40 years, and is one of sixteen albatross species identified as globally threatened in recent years, in large part due to drowning on fishery longlines. © Patricio Robles Gil/Sierra Madre

Lion-tailed macaque *Macaca silenus* (Endangered)

Named for the tuft of long hair at the end of its tail, this monkey could just as easily be referred to as the "lion-maned macaque," since many individuals —particularly the adult males— sport a prominent shag about the head and neck that evokes the distinctive headdress of Africa's largest feline. It is endemic to the mountainous Western Ghats of India, where it is believed to number several thousand individuals in highly-fragmented populations and is threatened by habitat loss. © Elio Della Ferrera

Rufous-necked hornbill *Aceros nipalensis* (Vulnerable)

This large Asian hornbill is now very rare across much of its extensive ancestral range in southern Asia and Indochina. An inhabitant of mature broad-leaved forests, it depends on large trees for nesting and feeding and vast tracts of forest for viable populations to survive. It is very susceptible to deforestation and habitat degradation and is also threatened by widespread hunting and trapping for food and for trade in live animals and their bills, used for ritualistic purposes. © M. Strange/Vireo

Leafy sea dragon *Phycodorus eques* (Data Deficient)

Resembling a piece of weed drifting in the water, the leafy sea dragon is one of only two sea-dragon species that are endemic to southern Australia. Named after the dragons of Chinese legend, these are bony fishes closely related to sea horses. As with the sea horses, it is the male sea dragons that carry the young (in this instance, in special brood pouches under the tail) until they hatch as fully-formed replicas of their parents. Sea dragons are poorly known, but there is concern that they are being affected by destruction and disturbance of their coastal marine habitat and by the growing interest to capture them for display in aquariums and use in oriental medicines. © Bill & Peter Boyle/Auscape

Hyacinth macaw *Anodorhynchus hyacinthinus* (Endangered)

The largest of the world's parrots and one of the most spectacular, the hyacinth macaw was heavily trapped for the live bird trade during the 1980s, when as many as 10 000 individuals are thought to have been taken from the wild. Although it is protected in the three South American countries where it occurs, illegal trade and habitat loss continue to put this species at risk. © Patricio Robles Gil/Sierra Madre

Chilean wood star *Eulidia yarrellii* (Endangered)

This hummingbird occurs only in two desert river valleys in northern Chile, where much of the land has been converted to cultivation. It inhabits small remnant patches of native scrub, and it is unclear to what degree it is dependent on native plants, which may be threatened by agricultural activities. © Luiz Claudio Marigo

Sambreeltjie (little umbrella) *Hessea undosa* (Vulnerable)

The Cape Province of South Africa is renowned for its large number of endemic species – found nowhere else in the world. Many of these are highly restricted in distribution, often confined to single mountaintops, and have very particular habitat requirements, which makes them easily threatened by human activities. This magnificent member of the Amaryllid family is one such species. Although the rocky terrain in which it grows currently limits the agricultural expansion that has caused it to decline, the development of new farming technologies and the move to crops that can grow in such rocky areas will always pose a risk. © Colin Paterson-Jones

Basking shark *Cetorhinus maximus* (Vulnerable; NE Atlantic subpopulation is Endangered)

The basking shark is a very large (up to 13.7 m in length) filter-feeder fish that inhabits cold-water seas in the Northern and Southern Hemispheres. Although widely distributed, it is only regularly seen in a small number of coastal locations and is probably never very abundant. It may not reach maturity until 18-20 years of age, may live to 50 years, probably does not "pup" every year, and the only known litter consisted of just 5 very large young. These traits render it extremely vulnerable to overfishing. The basking shark has been fished for centuries for its liver oil and meat, but more recently has been targeted for the international trade in shark fins. Basking shark fins are highly prized in this trade. © Jeremy Stafford-Deitsch/gerryellis.com

Radiated tortoise *Geochelone radiata* (Vulnerable)

Named for the star pattern of its carapace, the radiated tortoise is found only in the dry southernmost region of the island of Madagascar, where it is called "sokake." Although protected by local custom, it has been hunted by outsiders for food and curios, and captured for pets and the live-animal trade, but its relatively extensive range and locally high population densities have buffered it somewhat from these pressures. © Frans Lanting/Minden Pictures

Tibetan antelope *Pantholops hodgsoni* (Endangered)

The Tibetan antelope, or chiru, is native to the Tibetan Plateau of China and small areas of northern India and Nepal. These animals are killed illegally for their wool, which is known on international markets as "shahtoosh" or "king of wool." Shahtoosh is considered to be one of the finest animal fibers in the world, and since the 1980s, expensive shahtoosh shawls and scarves have been high-fashion status symbols in the West. Demand for shahtoosh has prompted an increase in poaching, and fueled a lucrative illegal trade which continues to thrive despite conservation and enforcement efforts by the Chinese government. As recently as 40-50 years ago, some 500 000-1 million Tibetan antelope may have roamed the Plateau, but now their numbers could be as low as 65 000-75 000. © George Schaller

Thick-billed parrot *Rhynchopsitta pachyrhyncha* (Endangered)

One of a few parrot species in the world that inhabit temperate climates, this species occupies conifer and pine-oak forests in the Sierra Madre Oriental and Occidental mountain ranges in Mexico. It no longer occurs in the U.S.A. Modification of its forest habitats and illegal trade have been the main agents of its decline. Attempts to reintroduce captive-bred animals to the wild have failed, owing to disease, behavioral deficits, and predation by birds of prey. © Patricio Robles Gil/Sierra Madre

Pitcher plant *Nepenthes rajah* (Endangered)

This carnivorous plant is found only on Mounts Kinabalu and Tambuyukon on the island of Borneo. It is one of the largest (and certainly the most famous) *Nepenthes* species, and has captured the interest of botanists and explorers since it was first discovered in 1858. The huge purple lower pitchers can hold up to 3 liters of water, and have been recorded catching rats in the wild. Overcollecting has been a problem, but most or all populations of this plant now occur in a protected area. © Phillip Cribb/RBG, Kew

Venus flytrap *Dionaea muscipula* (Vulnerable)

The Venus flytrap is found only in wetlands in the states of North and South Carolina in the United States. It traps insects between the two clam-like, hinged lobes on its leaf, which close quickly when the "trigger" hairs around them are touched. This plant has been heavily collected and has also lost tracts of its wetland habitat. Commercial collecting is now banned, and it is now widely cultivated, but concern for its habitat remains. © David Simpson/RBG, Kew

Southern cassowary *Casuarius casuarius* (Vulnerable)

Australia's largest land animal, all the cassowary has is the vestigial remains of wings. As it moves through the brush, it uses its hard casque for protection. The cassowary is an endangered species, with estimates of only 1 500 remaining (there may be fewer cassowaries in Australia than pandas in China). Its extinction could affect rainforest plant diversity as it helps spread the seeds of up to 100 tree and shrub species. © Jean-Paul Ferrero/Ardea London Ltd.

Chinese monal pheasant *Lophophorus lhuysii* (Vulnerable)

This colorful bird occurs in pockets over a relatively small area of montane scrublands and meadows in south-central China. It is thought to be declining, although not rapidly, due to degradation of its habitat from grazing herds of yak and disturbance from human activities such as hunting and collecting of herbs. © Kenneth W. Fink/Ardea London Ltd.

African elephant *Loxodonta africana* (Endangered)

Arguably the most totemic of the world's wild species, the African elephant has featured in the lore and culture of civilizations around the world for many centuries, in part owing to the extensive use of its tusks for ivory over this long period of time. Although there has been a serious decline in populations over the past half century, the African elephant still numbers around 500 000 animals across the African continent. © Patricio Robles Gil/Sierra Madre

Blue whale *Balaenoptera musculus* (Endangered)

The largest animal on the planet, the blue whale was heavily hunted —and its numbers severely depleted— in the last century. Dr. Randall Reeves, Chairman of the IUCN/SSC Cetacean Specialist Group, indicates that "there is still concern about the big blue whales of the Antarctic, and whether they can recover to any semblance of their former abundance —but there's not much that can be done except wait and see and make sure that they are not hunted for at least our lifetimes and probably those of our kids as well." © Mike Johnson

Polar bear *Ursus maritimus* (Lower Risk/conservation dependent)

According to Drs. Scott Schliebe and Stansilav Belikov, Cochairmen of the IUCN/SSC Polar Bear Specialist Group, until recently, it was generally thought that the long-term prognosis for the polar bear was favorable, but "our level of comfort has been shaken by recent studies of climate change and the occurrence in certain areas of the Arctic of higher levels of persistent organic pollutants. The habitat of polar bears once considered as pristine and stable can no longer be viewed in this context." © Kennan Ward

Giant sequoia *Sequoiadendron giganteum* (Vulnerable)

With a massive tapering trunk of up to 12 m in diameter and a maximum height of 95 m, the giant sequoia lays claim to being the world's largest tree. Like many other conifers, it is very long-lived: individual trees over 2 000 years old are not very rare, and some are known to be over 3 000 years old. First discovered by a hunter in 1852, the giant sequoia is endemic to California (U.S.A.), where it occurs in some 75 groves along a limited stretch of the western Sierra Nevada. It was commercially logged for timber until the 1950s and is now fully protected. Despite these measures, and its resistance to natural disasters and pathogens, few of the groves contain enough young trees to maintain future populations at their current levels. This is due at least in part to the suppression of fires, which has reduced reproduction. © Alan Watson/Forest Light

Grévy's zebra *Equus grevyi* (Endangered)

The largest of the seven wild horse species, Grévy's zebra occurs only in East Africa, in Ethiopia and Kenya, having gone extinct in Somalia. It is a grassland species, but extends into deserts where permanent water is available. Although it was hunted for its skins until the mid-1870s, current threats are habitat-related, and include competition from growing numbers of domestic livestock and extraction of water from natural water sources for human settlements and irrigation. © Kevin Schafer

Nassau grouper *Epinephelus striatus* (Endangered)

The Nassau grouper is a striking example of a once-common, widely distributed, commercially important marine fish that has become threatened with extinction. A popular food fish in the Caribbean, it, like many other groupers, gathers in large numbers for a few weeks each year to reproduce. These "spawning aggregations" make for easy fishing, and increasingly in recent years heavy fishing pressure has taken a serious toll: at least a third of all known spawning aggregations of this species have disappeared, and many others have dwindled from their former glory. Such is the concern for the Nassau grouper that the Bahamas government has formally protected several spawning aggregations in the hope of stemming declining numbers. © Bill Curtsinger

Rafflesia flower *Rafflesia pricei* (Vulnerable)

Named for the octogenarian amateur botanist who discovered it on Borneo in the 1960s, this is the showiest of the *Rafflesia* species, bizarre parasitic plants native to Southeast Asia. Living off the vines of plants in the grape family, Rafflesia produce no roots, stems or leaves, but only flowers and seeds. The flowers, which reach over 30 cm across, give off a smell of dead meat, which attracts the carrion flies that pollinate the plants. © Phillip Cribb/RBG, Kew

Scimitar-horned oryx *Oryx dammah* (Extinct in the Wild)

A victim of uncontrolled hunting and, to a lesser extent, competition with domestic livestock for scarce food resources, the scimitar-horned oryx is believed to have been exterminated in the wild in the early 1990s. It once ranged across several million square kilometers of semiarid Sahelian grassland and scrubland on both the northern and southern fringes of the Sahara Desert in northern Africa. The last known wild animals were recorded in Chad, where it is possible, though unlikely, that a handful still survive. Fortunately, the species breeds well in captivity —nearly 3 500 live in zoos around the world— and reintroduction efforts are under way in Morocco and Tunisia. © Nicolas Gaidet/Bios

Northern bald ibis *Geronticus eremita* (Critically Endangered)

This southern African species, with its glossy blue-green plumage, prominent red bill, and characteristic bald head, is more a grassland than a wetland bird. All the same, its habitat has declined as a result of commercial afforestation, intensive agriculture, mining, and other activities, and it is also hunted for use in traditional medicines and for ceremonial purposes. © Doug Wechsler

Kea *Nestor notabilis* (Vulnerable)

This olive-green parrot is found in wooded valleys and *Nothofagus* forests in the mountains of South Island of New Zealand. It also ventures into the subalpine scrub and alpine grassland to feed on seasonal fruit. It enjoys rolling in snow and becomes very bold and inquisitive around human habitation. Blamed with attacking sheep, it was shot under a bounty scheme in the 1940s, but is now partly protected. © Tui De Roy/The Roving Tortoise

Southern bluefin tuna *Thunnus maccoyii* (Critically Endangered)

With its cousin, the northern bluefin tuna, this is the largest bony fish in the world. It is also one of the most valuable: fresh bluefin tuna regularly sell for over US$30 000 a fish or US$50 per kilo in the Japanese fish markets that consume most of the world's catch. Its high value has led to serious overfishing and population declines, and efforts to manage the catch in the South Pacific are hindered by illegal fishing and other problems, including disagreements about how seriously at risk this fish truly is. © David B. Fleetham/Innerspace Visions

Sumatran rhino *Dicerorhinus sumatrensis* (Critically Endangered)

The most critically endangered of all five rhino species, the Sumatran rhino is also the least typical —small and hairy, in contrast to its more majestic and heavily-armored cousins. Only a few hundred of these rhinos are widely scattered in the rapidly disappearing rainforests of the Malay Peninsula and the islands of Borneo and Sumatra, where they are also threatened by poaching for their horns. © Günter Ziesler

Orangutan *Pongo pygmaeus* (Endangered)

The Indonesian word "orang-utan" means "man of the forest," which recognizes the close relationship between this impressive red-haired ape and our own species. There are two species of orangutan, this from the island of Borneo and the other, from the island of Sumatra. These are creatures of the trees, and their future is closely tied to that of primary forests remaining in Kalimantan, Sabah, and Sarawak on Borneo. Regrettably, these forests have suffered heavy damage in recent years as a result of widespread forest fires, logging, and their conversion to oil-palm plantations. At the same time, orangutans have continued to be hunted for food and captured for sale as pets, which has further reduced their numbers. © Günter Ziesler

Cocha (lagoon) in Manú National Park, Peru

One of the world's best-known rainforest parks, Manú harbors a huge diversity of plant and animal life. However, on one side of the park lies the Camisea gas field, one of the world's largest undeveloped gas fields, and on the other, illegal and uncontrolled gold mining activities are destroying habitats and polluting waterways. The long-term integrity of this natural treasure will depend on sensitive energy planning and management of extractive activities. © Patricio Robles Gil/Sierra Madre

Przewalski's horse or takhi *Equus ferus przewalski* (Extinct in the Wild)

A close relative of the domestic horse, the takhi once roamed widely over the steppes of central Asia, China, and western Europe, but was last seen in the wild in 1966. A population of over 1 000 takhi in captive herds is being managed through a collaborative effort, thus securing it from extinction for the foreseeable future. Successful reintroduction projects are taking place in Mongolia and China. © Art Wolfe

Brown bear *Ursus arctos* (Lower Risk/least concern)
The most widespread of the world's eight bear species —formerly occurring across much of the Northern Hemisphere—, the brown bear is now gone from much of its former range and in many areas persists only in small pockets. In Spain, for example, it is now found only in the Cantabrian Mountains, in two subpopulations totaling fewer than 100 animals. The main threats are habitat loss and overhunting. © Francisco Márquez

Western Australian swamp turtle *Pseudemydura umbrina* (Critically Endangered)
One of the world's rarest freshwater turtles, the Western Australian swamp turtle inhabits winter-wet swamps in a region that has been almost completely converted to agriculture or urban development. It was protected in the 1960s in two small nature reserves, but declined to fewer than 50 turtles by the 1980s. A recovery program has succeeded in increasing its numbers to 130 in the wild and 200 in captivity. © Jiri Lochman/Lochman Transparencies

Tropical rainforest, Guatemala
The majesty of the ancient Mayan city of Tikal, first settled around 700 BC, competes today with the surrounding tropical rainforest. Towering buttresses rise up to the jungle's green canopy, where howler monkeys swing effortlessly through the branches of ancient trees, and the rich smells of earth and vegetation assault the senses, along with warbling and other songs of unseen birds and the buzz of insects and tree frogs. These forests are now threatened by human encroachment. © Pablo Cervantes/Sierra Madre

Apollo butterfly *Parnassius apollo* (Vulnerable)
This large butterfly inhabits montane areas in northern Eurasia, extending as far as Mongolia and China. More than 160 different subspecies have been described, along with specific regional populations; a number of these are extinct or endangered. Its range has declined, and populations are often isolated. The causes for its decline are not properly understood, but are thought to include roadways, conifer plantations, climate change, and acid rain. © Francisco Márquez

Tomato frog *Dyscophus antongilii* (Vulnerable)
One of over 170 amphibian species found only in Madagascar, the tomato frog occurs in wetland habitats along the east and northeast coasts of the island. Its spectacular coloring and relatively large size have made it an appealing subject for pet-keepers and amateur herpetologists, which fueled exports of this otherwise little-known frog. It is now protected from international trade, but long-term habitat degradation and destruction continue to pose a threat to this species. © Michael Durham/gerryellis.com

Blue crane *Grus paradisea* (Vulnerable)
The national bird of South Africa, where almost all of its representatives occur, the blue crane is a graceful bird with flowing bluish-gray plumage sweeping to the ground. It is primarily a grassland species and uses wetlands only occasionally. Its overall numbers have declined drastically in recent decades, the main causes being widespread poisoning on agricultural land and afforestation of large tracts of its grassland habitat. © Michael & Patricia Fogden

Roraima's tepui, Venezuela

Lying on Venezuela's border with Guyana and Brazil, the massive table-top mountain or "tepui" of Roraima, which peaks at 2 810 m, is one of the highest in South America. Its name, which in Pémon means "the great, ever-fruitful, mother of streams," evokes the drama and mystery of this strange site dominated by black rocks, shrouded in fog, and coated with lichens and mosses forming colored rock-pools adorned with beautiful flowers and odd-looking plants found nowhere else in the world. The plants of the surface of the tepuis have evolved in total isolation: over 2 000 plant species are endemic to this alien rock landscape. © Patricio Robles Gil/Sierra Madre

Grant's golden mole *Eremitalpa granti* (Vulnerable)

As Dr. Edwin Gould has stated in the IUCN/SSC Action Plan for African Insectivora and Elephant-Shrews: "Extinction of an elephant species will resound around the world; extinction of an insectivore or elephant-shrew will go undetected for decades, and even then few people will care. If natural habitats are to be destroyed, at the very least let us consider these precious few areas that are inhabited by the beautiful animals we have learned to treasure for their own sake." © Michael & Patricia Fogden

Spectacled bear *Tremarctos ornatus* (Vulnerable)

Of the world's eight bear species, the spectacled bear is the only one that occurs in South America. Named for the white markings on its face, which often encircle one or both eyes, it is widely but patchily distributed along the Andes range, where it is threatened by a number of factors, in particular loss of habitat, through logging and conversion to agriculture and human settlement, and hunting. As is the case with bears in other cultures, the spectacled bear had enormous spiritual and ritualistic importance in pre-Columbian times, some of which persists to this day. According to Dr. Bernard Peyton, who has long championed the conservation of this species, "a lot of hope for self-improvement will die with the extinction of spectacled bears in the wild." © Patricio Robles Gil/Sierra Madre

Bardick *Echiopsis curta* (Vulnerable)

Some of the world's most infamous snakes belong to the family Elapidae, a group of venomous snakes that include the cobras, mambas, and this species, the bardick. Endemic to Australia, the only continent harboring more venomous than non-venomous snakes, the bardick occurs in southern shrubland habitat that is declining as a result of overgrazing by livestock, clearance for grazing and agriculture and, possibly, an inappropriate fire regime. © Jiri Lochman/Lochman Transparencies

Silversword *Argyroxiphium sandwicense* (Vulnerable)

One of the most striking of Hawaii's endemic plants, the magnificent silversword is found only at high elevations on Mauna Kea on Hawaii and Haleakala on Maui. Its attractiveness was a significant factor in its early decline —climbers on the mountains often dug up and brought back plants as proof that they had reached the summit. Browsing by goats and cattle further depleted populations, to such an extent that concern was expressed for its survival as early as the 1930s. Protection from vandalism and fencing to keep away grazing animals have led to a dramatic recovery, at least on Haleakala, where there are many more plants than there were in 1935. © Frans Lanting

Iiwi *Vestiaria coccinea* (Lower Risk/near threatened)

The scarlet iiwi is the most populous of Hawaii's honeycreeper species, all of which are unique to this U.S. state. Nonetheless, its status does not mean its survival is certain: more than half of the state's 30 honeycreeper species are now extinct. Although it was once used extensively in native Hawaiian dress, avian malaria may be responsible for more recent disappearances of this species from mid-elevation forests. © Frans Lanting

Jamaican iguana *Cyclura collei* (Critically Endangered)

Regarded as the rarest lizard in the world, this iguana was believed extinct until the rediscovery, a decade ago, of a tiny remnant population in the Hellshire Hills of Jamaica. The wild population of 100-200 animals is threatened by predation by introduced mongooses and encroachment into its habitat, but a successful "headstarting" program, involving the release of animals that have been raised in captivity and other measures, guided by the IUCN/SSC Iguana Specialist Group, offer hope for this species' future. © Glenn Gerber/IUCN Iguana Specialist Group

African wild dog *Lycaon pictus* (Endangered)

With its oversized round ears, mottled coat, and bushy white tail, the African wild dog seems an endearing creature. However, as one of the most efficient mammalian hunters in the world, it is persecuted by humans over large parts of its range. Wild dogs live in packs formed of males from one pack and females from another pack. In most instances, the dominant male and female (the "alpha pair") are the only ones to breed, and the other dogs help to feed the large litter of pups and guard them when the rest of the pack is out hunting. Outside of their three-month "denning" period, wild dogs are great wanderers whose hunting range may cover as much as 2 000 km^2. © Lex Hes

Tropical lady's slipper orchid *Paphiopedilum sanderianum* (Endangered)

With its long, dangling petals that reach a meter or more in length, *P. sanderianum* is one of the most striking of all orchids. It has very specific habitat requirements, which may explain its rarity. Its extremely long petals are thought to attract hover flies, almost certainly its potential pollinators. Most of the colonies of this fine orchid are found within the confines of a national park. However, the closely-guarded secret of its habitat has been discovered by commercial orchid collectors who have reportedly stripped large numbers of plants from the wild for sale at high prices. © Phillip Cribb/RBG, Kew

Jade slipper orchid *Paphiopedilum malipoense* (Endangered)

This recently described (1984) lady's slipper orchid is well known for the tall inflorescence that bears only one large, raspberry-scented flower. It occurs only in karst limestone in southwestern China and across the border in similar habitats in northern Viet Nam. Wild-collected plants are often sold in flower markets in the region, where the price this species now fetches suggests that it has become a very rare plant in the wild, possibly the rarest of the Chinese lady's slipper orchids. © Phillip Cribb/RBG, Kew

Iberian lynx *Lynx pardinus* (Critically Endangered)
The most threatened of all the wildcats, the Iberian lynx numbers only about 600 animals, most of them in Spain, in heavily fragmented populations, only two of which are considered large enough to survive over the long term. Once widespread in the mountain woodlands and open pastures of Spain and Portugal, it has lost ground to agriculture and development, and to the decline of its main prey, the European rabbit, due to disease. © Jorge Sierra

Wattled crane *Grus carunculatus* (Vulnerable)
This large crane occurs in several isolated small populations in Africa and over a large area of wetlands in central southern Africa, in Botswana, Zambia, and Zimbabwe. The primary threat to its survival is loss of this wetland habitat to intensified agriculture, draining, rice cultivation, and flooding by dams, as well as other forms of disturbance and degradation, including persecution by humans and hunting. © Theo Allofs

Andean flamingo *Phoenicoparrus andinus* (Vulnerable)
The Andean flamingo inhabits alkaline and salt lakes on the high Andean plateaus of Peru, Chile, Bolivia, and Argentina. Its numbers were severely reduced in the mid-twentieth century by collection of eggs for sale as food, and its current population of some 34 000 birds continues to decline as a result of the indirect effects of mining activities, unfavorable water levels, and human disturbance, as well as low-level hunting for food and feathers. © Günter Ziesler

European bison *Bison bonasus* (Endangered)
The European bison is a relic of ancient times. Once distributed throughout western, central, and southeastern Europe, it began to decline due to intense hunting pressures in the eighth century until it was nearly exterminated eleven centuries later. By the early twentieth century, only two free-ranging populations remained: one in eastern Poland, which declined rapidly from 785 individuals in 1915 and became extinct after World War I (April 1919), and the other in the northwestern Caucasus region, which met the same fate in 1927. The only survivors were 54 bison held in European zoos. All European bison currently in existence are descended from just 13 of these zoo animals. © Cyril Ruoso/Bios

Gredos ibex *Capra pyrenaica victoriae* (Vulnerable)
This male ibex, characterized by its enormous, back-curving horns, longingly woos a female. Ibex inhabit craggy terrain between the timber and snow lines of the Iberian Peninsula. The species dwindled steadily over the past century as a result of poaching and environmental factors, in some cases natural disasters like landslides. Now extinct in the Pyrenees, the species is still present in the Sierra de Gredos. Although protected, factors such as frequent fires, logging of native woodlands, and domestic and invasive alien species still threaten its survival. © John Cancalosi

Steller's sea eagle *Haliaeetus pelagicus* (Vulnerable)
This northeast Asian raptor is confined to narrow strips of coast along the Okhotsk and Bering seas, extending inland to forested valleys along the lower reaches of the salmon-rich rivers that flow into them. In Russia, it is threatened by habitat alteration from hydroelectric power projects, coastal development for the petrochemical industry, and logging. Industrial pollution and overfishing are additional threats. © Konrad Wothe